international®
AIR POWER
REVIEW

AIRtime Publishing

international
AIR POWER
REVIEW

Published quarterly by AIRtime Publishing Inc.
191 Post Road West, Westport, CT 06880
Tel (203) 454-4773 • Fax (203) 226-5967

© 2006 AIRtime Publishing Inc.
V-22 cutaway © Mike Badrocke

Photos and other illustrations are the copyright of their respective owners

ISBN: 1-880588-89-7 (casebound)
Softbound Edition ISSN 1473-9917

Publisher
Mel Williams

Editor
David Donald   e-mail: dkdonald@hotmail.co.uk

US Desk
Tom Kaminski

Russia/CIS Desk
Piotr Butowski, Zaur Eylanbekov

Europe and Rest of World Desk
John Fricker, Jon Lake, Marnix Sap •

Correspondents
Argentina: Santiago Rivas and Juan Carlos Cicalesi
Australia: Nigel Pittaway
Belgium: Dirk Lamarque
Brazil: Claudio Lucchesi
Bulgaria: Alexander Mladenov
Canada: Jeff Rankin-Lowe
France: Henri-Pierre Grolleau
Germany: Frank Vetter, Bernd Vetter
India: Pushpindar Singh
Israel: Shlomo Aloni
Italy: Luigino Caliaro
Japan: Yoshitomo Aoki
Netherlands: Tieme Festner
Romania: Danut Vlad
Spain: Salvador Mafé Huertas
USA: Rick Burgess, Brad Elward, Mark Farmer (North Pacific
    region), Peter Mersky, Bill Sweetman

Artists
Mike Badrocke, Piotr Butowski, Tom Cooper, Chris Davey

Controller
Linda DeAngelis

Sales Director
Jill Brooks

Origination by Chroma Graphics, Singapore
Printed in Singapore by KHL Printing

All rights reserved. No part of this publication may be copied, reproduced, stored electronically or transmitted in any manner or in any form whatsoever without the written permission of the publishers and copyright holders.

International Air Power Review is published quarterly in two editions (Softbound and Deluxe Casebound) and is available by subscription or as single volumes. Please see details opposite.

Acknowledgments
John Haire (AFFTC PAO) and Lt Col Dwayne Opella, 410th FLTS operations officer for their help with the YF-117 feature.
Mr John Milliman, NAVAIR Public Affairs – PEO(A), and Mr Ed Steadham, Sikorsky Aircraft, for their assistance with the CH-53K Debrief feature.
Heidi Wendt and Linn Lichtermann for their help with the Saab UAV Debrief feature.
AVM (Ret) Norm Gray, Deputy CEO DMO, Patrick Gill, Vice President AEW&C Programmes, Boeing, Ken Morton, Communication Director Australia/NZ, Boeing and Dave Sloan, Air Force Systems, Boeing, for their assistance with the Wedgetail report.
The 1st Fighter Wing for its support with the F-22 article. Special thanks to 1st FW Public Affairs Officer Lt Natasha Waggoner, Lt Colonel Wade Tolliver, SSGT Thomas Doscher, Lt Elizabeth Kreft and Brig. General Burton Field. Special thanks to USAF Reserve PAO, Colonel Audrey Bahler, and the 459th ARW, as well PAO Lt. Tim Smith.
James Darcy, V-22 Osprey Public Affairs; Col. Bill Taylor, UMSC, Program Manager, V-22 Joint Program Office; Lt. Col. Kevin Gross, USMC; Mike Anderson, Bell Boeing V-22 Joint Program Office Vice President and Deputy Program Manager; Bob Leder, Bell Helicopter Textron Public Affairs; Madonna Walsh, Boeing Public Affairs for their help with the V-22 feature.
Cols Randy Ball, Matthew Erichsen, Lt Cols Thomas Bussiere, Doug Hart, Mike Lenahan, Greg Miller, Paul Tibbets IV, Tom Pritchard, Majs Joe Brooks, Dwayne Cowles, Brian Gallo, Mike Means, Capts Ryan Bailey, Jason Ceminsky, Brenda Hendricksen, Jennifer Jeffords, Jared Kennish, Lts Rick Alvarez, Mary Olsen, SMSgt Jim Blucher, MSgts Gary Howard, TSgts Dennis Angier, Jeff Capenos, Matt Rose, SSgt Joe Prouse and all the others that helped with the 509th BW article.

Contact and Ordering Information
(hours: 10 am - 4.30 pm EST, Mon-Fri)
addresses, telephone and fax numbers
International Air Power Review, P.O. Box 5074, Westport, CT 06881, USA
    Tel (203) 454-4773 • Fax (203) 226-5967
    Toll free within USA and Canada: 1 800 359-3003
    Toll free from the United Kingdom, Belgium, Denmark, France, Germany,
        Holland, Ireland, Italy, Luxembourg, Norway, Portugal, Sweden and
        Switzerland (5-6 hours ahead): 00 800 7573-7573
    Toll free from Finland (6 hours ahead): 990 800 7573-7573
    Toll free from Australia (13-15 hours ahead): 0011 800 7573-7573
    Toll free from New Zealand (17 hours ahead): 00 800 7573-7573
    Toll free from Japan (14 hours ahead): 001 800 7573-7573

website
    www.airtimepublishing.com
e-mails
    airpower@airtimepublishing.com
    inquiries@airtimepublishing.com

Subscription & Back Volume Rates
**One-year subscription (4 quarterly volumes),
inclusive of ship. & hdlg./post. & pack.:**

Softbound Edition
USA $59.95, UK £49.50, W. Europe EUR 88, E. Europe EUR 92,
Canada Cdn $99, Rest of World US $112 (air)

Deluxe Casebound Edition
USA $79.95, UK £69.95, W. Europe EUR 120, E. Europe EUR 128,
Canada Cdn $132, Rest of World US $154 (air)

**Two-year subscription (8 quarterly volumes),
inclusive of ship. & hdlg./post. & pack.:**

Softbound Edition
USA $112, UK £94, W. Europe EUR 169, E. Europe EUR 177,
Canada Cdn $187, Rest of World US $219 (air)

Deluxe Casebound Edition
USA $149, UK £132, W. Europe EUR 230, E. Europe EUR 246,
Canada Cdn $247, Rest of World US $299 (air)

Single/back volumes by mail (each):
Softbound Edition
USA $19.95, UK £14.95, W. Europe EUR 24, E. Europe EUR 25,
Canada Cdn $32, Rest of World US $32
(incl. ship. & hdlg./post. & pack.)

Deluxe Casebound Edition
US $24.95, UK £18.95, W. Europe EUR 32, E. Europe EUR 34,
Canada Cdn $38, Rest of World US $42
(incl. ship. & hdlg./post. & pack.)

Prices are subject to change without notice. Canadian residents please add GST. Connecticut residents please add sales tax.

Volume Nineteen

CONTENTS

MAJOR FEATURES PLANNED FOR VOLUME TWENTY

Focus Aircraft: Eurofighter Typhoon, **Warplane Classic:** Boeing C-97/KC-97, **Variant Briefing:** Martin B-57 Canberra, **Air Power Analysis:** Germany Part 1, **Special Report:** Lockheed Martin F-117, **Air Combat:** RAF nuclear forces, **Technical Briefing:** Mil Mi-8 upgrades

PROGRAMME UPDATE

Eurofighter Typhoon

The long-awaited mid-December 2005 signature of Euro 16 billion ($21 billion) contracts for Tranche 2 orders of 236 extended-role Eurofighter Typhoons by the four partner nations of Germany, Italy, Spain and the UK, followed NETMA's Batch 2 Type Acceptance certification on 13 December. This included, for the first time, flight-operation clearance of the Typhoon's Defensive Aids Sub-System (DASS) and Multiple Information Distribution System (MIDS) datalink. Also cleared were air-to-air refuelling and external fuel tank carriage, launch of medium-range Raytheon AIM-120 AMRAAM air-to-air missiles, and the Typhoon's innovative Direct Voice Input (DVI) control system, to reduce pilot work-loads. Following initial Batch 1 Type Acceptance on 30 June 2003, two-seat series production aircraft have now been delivered to each of the four air forces, and service entry training is proceeding. This follows initial flights of three twin-seat Instrumented Production Aircraft (IPA) in April 2002, and the first flight of a similar single-seat Typhoon (IPA4), on 26 February 2004.

In London, Defence Secretary Geoff Hoon said in December that Tranche 2 contracts will provide the RAF with an additional 89 aircraft costing some £4.3 billion ($8.17 billion), representing a programme unit cost of around £48.3 million ($91.8 million). With manufacturing plants in Germany and Spain, EADS is receiving Tranche 2 Typhoon contracts worth Euro 4.3 billion ($5.84 billion).

Associated weapons systems will be extended to include precision-guided missiles for additional limited ground-attack roles, for which RAF Typhoons will use Raytheon Paveway IVs with a 500-lb (227-kg) GBU-12-based Mk 82 warhead, and Litening targeting pods. Tranche 2 Eurojet EJ200 turbofans will also incorporate new digital engine control and monitoring units (DECMUs). As a 36 per cent shareholder in Eurojet, the Rolls-Royce share of Tranche 2 component production for 519 EJ200 turbofans is worth more than £750 million ($1.425 billion). Rolls is now also starting to assemble, test and deliver 195 engines for the RAF at about 40 annually until 2010.

Except for the UK, other Eurofighter partners' Typhoons will include IRIS-T close-combat advanced AAMs in their weapons fit, following the 20 December signature by the German Federal Office for Military Technology and Procurement (BWB), with Nuremberg-based Diehl BGT Defence company, of a Euro 1 billion ($1.36 billion) series production contract. Diehl is prime contractor in the six-nation IRIS-T programme, which Germany leads, in succession to original Sidewinder AAM production in Europe under US licence. IRIS-T will be operated by the Eurofighters, Tornados, GD F-16s, Boeing/MDC F/A-18s and JAS 39 Gripens in the air forces of Germany, Greece, Italy, Norway, Spain and Sweden, as programme partners, from production of more than 4,000 missiles by 2011.

Eurofighter Typhoon s/n IS002, which made its initial flight in December at Alenia Aeronautica's Caselle plant near Turin, is the second single-seat Italian-built version, although the first scheduled for AMI operational service. Flown by Alenia Defence chief test-pilot Commander Maurizio Cheli, IS002 successfully completed its basic tests during its initial 65-minute sortie. After further trials, it will join the Typhoons so far delivered to the AMI that are replacing the Lockheed F-104S ASA-M Starfighters of the 4° Stormo at Grosseto.

Italy will receive 19 single- and 10 two-seat Typhoons from Tranche 1 production, followed by 43/3 from the Tranche 2 contract, signed on 14 December. AMI Typhoon requirements are due for completion by 44/2 Typhoons in Tranche 3, if this survives increasingly stringent military budget economies in the four Eurofighter partner nations.

Full day/night air-refuelling clearance for the Typhoon was achieved late last year, after eight successful trial sorties completed by Alenia Aeronautica with the instrumented production IPA2 aircraft, flying from the AMI's Practica di

PROJECT DEVELOPMENT

Argentina

AT-63 programme resumed

Roll-out on 15 December 2005 of the first upgraded AT-63 Pampa armed jet-trainer at Lockheed Martin Aircraft Argentina SA's (LMAASA) Cordoba plant followed reactivation of this original $230m year-2000 programme, suspended in 2003 through defence budget economies. Two five-year contracts awarded by the Argentine Government to LMAASA in February 2004 revived earlier plans to upgrade 12 existing Argentine air force (FAA) IA 63 Pampa intermediate jet-trainers as AT-63s, with a 3,500-lb (15.57-kN) thrust Honeywell TFE731-2C turbofan, 1553B digital avionics, a glass cockpit, HOTAS, and five stores pylons, for additional weapons training and light ground-attack capabilities. Also included is production of another six new AT-63s for the FAA, and six more for international sales, plus maintenance and modification services for air force aircraft, engines and accessories.

Originally developed with assistance from Dornier Apha Jet trainer inputs, the new-generation Pampa is claimed to combine ease of maintenance and airframe stability with a modern Mil Std 1553B databus avionics suite, sophisticated mission computers, and advanced multi-function cockpit and head-up displays. These are integrated with advanced weapon systems, while achieving significant reductions in maintenance costs by using modular approaches to systems design. Ground testing is now progressing, prior to flight-trials, expected to start mid-2005.

One of the FAA's dozen IA-63s – actually the first prototype – was lost, together with its pilot, in a crash on 12 December, near Punta Indio air base in the Buenos Aires region.

France

Storm Shadow trials completed

A highly successful firing of MBDA's Storm Shadow/SCALP EG long-range cruise missile was completed by an Aéronavale Rafale MO2 combat aircraft flying from *Charles de Gaulle* in the Atlantic on 3 December. Released from the Rafale's centre-line pylon at 20,000 ft (6096 m) and Mach 0.8, the missile followed its pre-programmed flight trajectory over the sea and land areas of the CEL (Centre d'Essais des Landes) range towards its designated target.

During its cruise phase, the Storm Shadow was guided by its INS/GPS and infra-red imaging terrain profile-matching combined navigation system, before hitting the target with extreme precision. The difference between intended and actual target impact points proved to be lower than the missile system's specified metric accuracy tolerance.

This sixth Storm Shadow test success marks the end of the Rafale integration programme, and concludes a highly successful year for MBDA's cruise missile systems, which also included an APACHE launch, from the Rafale and Mirage 2000. Storm Shadow qualifications and successful combat operation have further been achieved for RAF Tornado GR.Mk 4s, and over 600 missiles have already been delivered to various customers, including Italy and Greece.

Greece

First HAF C-27J flies

Flight clearance of the first production Alenia/Lockheed Martin C-27J Spartan twin-turboprop transport on order for the Hellenic air force (HAF) started in mid-December with an initial sortie from Alenia Aeronautica's Turin-Caselle plant. Crewed by Commanders Agostino Frediani, Gianluca Evangelisti and Mario Mutti, the C-27J's first 83-minute flight initiated a successful contractual time

On 19 February 2006 the first Lockheed Martin F-35 Joint Strike Fighter ventured outside for the first time. Its assembly had been completed two days before, and it was being moved to a fuelling facility for a thorough fuel system check-out. Structural coupling and ground-vibration testing will follow. Engine runs will begin in late spring and will lead into taxi tests before the first flight.

Mare air base, near Rome. The initial night sortie concentrated mainly on operating with only the position lights from the refuelling AMI Boeing KC-707, and its trailing basket. The tests also validated the design decision to exclude illumination of the refuelling probe, which presented no difficulties in completing the tests.

In January 4° Stormo Typhoons became the first of the breed to be placed into operational service, standing air defence alert. They were used to cover the Winter Olympics held in and around Turin during February.

Recent German press claims that Luftwaffe Eurofighters have only limited combat readiness after soaring R&D costs were rejected in January by the German Defence Ministry, which confirmed that all development tasks were running as planned. Commenting on alleged mission capability limitations, it was emphasised again that in accordance with normal aerospace industry practice, Eurofighter performance standards are implemented and released in stages.

The Ministry said that multi-national Development Aircraft have already proved Eurofighter's capabilities, including gun-firing, air-refuelling, carriage of external fuel tanks, and air-to-air missile launches. Series production aircraft have all of these capabilities, but release of relevant clearance to service pilots is a progressive, staged approach.

Eight two-seat Eurofighters were then also in Luftwaffe service, one being used exclusively for maintenance training. The others are currently being operated in evaluations of operational issues, such as tactics and logistics, and their service release will gradually be upgraded to full production standards. It was not until 14 February,

however, that EADS Military Aircraft delivered the first single-seat German series production Typhoon to the Luftwaffe.

By the end of February Eurofighter had delivered 78 production aircraft, including the five IPAs. The UK had received 25 operational aircraft, Germany 21, Italy 14 and Spain 13. The combined test/customer fleet passed 10,000 hours in November 2005. Austria is due to receive the first of its 18 aircraft in May 2007, and Eurofighter has achieved a major sale of the Typhoon to Saudi Arabia.

---

schedule, to start deliveries by late January.

Currently, Alenia Aeronautica is building 16 C-27Js at its Pomigliano d'Arco (near Naples) and Caselle factories, to fulfil orders for 12 each for the Greek and Italian air forces. Interest in the C-27J has also been reported in the US, where the Army has initiated a joint initial fixed-wing twin-turboprop aircraft requirement through the JCA programme, for replacement of 43 National Guard Short C-23 Sherpas. The ANG has indicated the Italian/American aircraft as the likely Sherpa successor, with an initial requirement for 37 aircraft.

Following an earlier North American demonstration tour, interest in the C-27J is also being expressed by Canada, which needs 15 replacements for its SAR DHC-5 Buffalos and Lockheed C-130s. The C-27J has also been formally evaluated by the air forces of Australia, Taiwan, Ireland, Portugal, Bulgaria, Czech Republic, and other recent NATO member countries.

Greece joins M-346 programme

The Hellenic aerospace industry is planned to become a primary partner with Finmeccanica's Aermacchi company for production of the latter's M-346 advanced transonic twin-turbofan lead-in fighter-trainer (LIFT) trainer, from a Memorandum of Understanding (MoU) signed by the Greek Defence Ministry in January. Currently in its early industrialisation phase, with the first of two prototypes flying since 15 July 2004, continuing towards type certification, the M-346 programme is fully supported by the Italian Government.

India

New AEW&C programme

As expected, EMBRAER's announcement in February of a Memorandum Of

Understanding (MOU) signed with the Indian DRDO (Defence R&D Organisation), to support a new Airborne Early Warning & Control (AEW&C) development programme using indigenous radar, could involve initial IAF acquisition of another three EMB-145 Legacy twin-turbofan transports as the associated platforms. These would follow 2003 Delhi government orders for five EMB-145s, including four for IAF operation on VIP and official transport roles, and the fifth by India's Border Security Force. Equipped with defensive aids sub-systems, including chaff and flare dispensers, the new EMB-145s will reportedly begin to arrive in India within a few months.

Last year, the Indian government cleared proposals from the Defence Research and Development Organisation (DRDO), to develop and integrate the new AEW&C system with indigenous radars, for a total programme cost expected to total Rs18bn ($415m). This is a continuation of a long-term DRDO airborne radar development programme, temporarily halted by the loss of a HAL/Avro 748 test-bed, fitted with a large dorsal radome, and its crew of eight pilots and technicians, on 11 January 1999.

In India, the new aircraft will supplement an initial three Russian-supplied Il-76-based Beriev A-50EhI AEW&C aircraft costing $1.1bn, with Israeli Elta EL/M-2075 Phalcon phased-array radar in a fixed dome. IAF deliveries will start in 2006, with up to five received by 2008.

International

A400 construction begins

Production of the first major airframe component for the prototype A400M military transport was formally launched on 26 January, at the Varel facility of Airbus Germany. Representatives of the seven

customer nations joined government officials and members of the Airbus Military management team to witness milling of the first lower-fuselage frame. One of 18 similar frames located in the centre-fuselage section, the 5.4-m (17-ft 9-in) long element is milled down to 25 kg (55.1 lb) from a single aluminium billet of over two tonnes. All the machined waste is being re-cycled.

Enforcing the rigorous programme timetable, the first metal cut for the A400M comes only 18 months after programme launch, in May 2003. A400M first flight is planned in 2008, with deliveries of 180 beginning from 2009 to seven European NATO nations, comprising Belgium (7); Germany (60); France (50); Luxembourg (1); Spain (27); Turkey (10); and UK (25).

Russia

Fifth-generation fighter for 2007

In a recent Moscow press interview, RFAF commander-in-chief General of the Army Vladimir Mikhailov claimed that flight development of the prototype of Russia's PAK FA fifth-generation fighter was planned in 2007. He said that Sukhoi computer modelling and design studies for the "completely new" PAK FA were well advanced, and in a recent visit to the company he had told General Director Mikhail Aslanovich Pogosyan he had no objections to the first flight being on 31 December, providing it was in 2007.

Gen. Mikhailov said that several problems remained unresolved, particularly concerning armament and avionics, as well as the limited finance available for defence over the past few years, indicating the possibility of sharing development costs through an international project. He also confirmed planned Defence Ministry procurement of three twin-Perm PS-90A-engined Tupolev Tu-214 transports

required for long-range international air travel, and equipped with the latest avionics to meet new ICAO commercial general air traffic management (GATM) standards introduced on 1 January.

Having insisted on rectification of some deficiencies in the ZMKB Progress/Ivchenko D-27 prop-fan engines, and completion of more ground-tests, Gen. Mikhailov said that Russia was now prepared to continue flight-development with the Ukraine of the Antonov An-70 heavy-lift military transport.

South Africa

First Denel-assembled Hawk

The first of 23 BAE Systems Hawk Mk 120 lead-in fighter trainers to be assembled in South Africa by Denel Aviation made its initial flight on 13 January from Johannesburg International Airport. It followed the sole UK-assembled Hawk in the SAAF contract, which has been in South Africa for some time, and is currently based at the Air Force's Test Flight & Development Centre near Bredasdorp. It is now completing flight-tests of the navigation and combat training systems, designed and developed by the local Advanced Technologies and Engineering (ATE) company.

Denel's first Hawk was initially flown by South African industry test-pilot Dave Stock, together with BAE Systems test-pilot Gordon McClymont, on an 80-minute sortie during which the aircraft's flight controls, handling response and essential systems functionality were successfully confirmed. Denel, which manufactures Hawk aerostructures, including the tailplane and air-brake, is assembling 23 of the 24 new jet-trainers. SAAF Hawk deliveries are on schedule to start by mid-year, continuing at a rate of two aircraft per month until mid-2006.

Saab Gripen

Right: On 13 December 2005 Saab's trials Gripen 39101 carried a functional subsystems-equipped Meteor missile into the air for the first time. The first launch will take place in 2006 at Vidsel.

Below: A single- and two-seat Gripen are seen conducting refuelling trials with a Swedish Air Force C-130 in November 2005.

Below right: With Saab test pilot Magnus Olsson and South African Charl Coetzee at the controls, the first Gripen for the SAAF made its maiden flight from Linköping on 11 November. The aircraft is to transfer to Bredasdorp for trials in South Africa.

SAAF C-in-C Lt Gen. Roelf Beukes recently visited the Hawk final assembly line, and congratulated Denel and BAE Systems on their outstanding work progress. BAE Systems vice-president for South Africa Jonathan Walton paid tribute to the successful close collaboration and valuable contributions from the SAAF, Armscor, Denel, ATE and Rolls-Royce, in the Hawk programme, which was also delivering $8.7bn of new economic benefits to the country in the defence, aerospace and numerous civil sectors.

South Africa joins A400M

The Airbus Military A400M multi-role mission transport aircraft design and production programme gained an additional partner in December, when the South African government announced that it has accepted an invitation to participate in this already multi-national project. South Africa's planned membership is accompanied by a commitment to fund procurement of eight to 14 A400Ms as the programme matures between 2010 and 2014. The cost of eight aircraft is quoted as R837m ($108.43m).

Negotiations are in progress with Airbus Military to determine the terms of agreement for the South African government's participation. Denel and Aerosud are also discussing with Airbus Military details of industrial partnership contracts around specific dedicated work-share packages that will become effective following pending agreements reached between government and Airbus Military. Previously, 180 A400Ms had been ordered in May 2003 by seven European NATO nations, to fly from 2008, with deliveries starting from 2009.

The SAAF's current military transport and airlift capability is achieved by nine C-130B Hercules dating back to 1959, and

now approaching the end of their working lives. They will need replacing by about 2010, and already over the past three years, South Africa has had to spend over R100m ($16.3m) for privately-owned airlift capabilities to deploy personnel, resources and materiel into certain African peace-keeping operations. The A400M will allow South Africa to upgrade its airlift capability, and assist in strengthening its aeronautical industry.

South Korea

KAI/Eurocopter partnership

Korea has announced its selection of Eurocopter as the primary partner of Korea Aerospace Industries (KAI) in a new programme under which the two companies will jointly develop the nation's first indigenous military transport helicopter. Known as the Korean Helicopter Program (KHP), the new aircraft will replace the Army's fleet of transport and liaison helicopters. The KHP's six-year development phase will run from 2006 to 2011 and 245 helicopters will be built under the 10-year production that follows. Capable of carrying a crew of two and 11 troops, the KHP is expected to cost $6-8 billion. Eurocopter's stake in the development programme amounts to 30 per cent, whereas its share of the production will be 20 per cent. KAI and Eurocopter have agreed to establish a 50/50 subsidiary that will market the export version of the KHP.

United States

Joint Strike Fighter update

Lockheed Martin completed assembly of the major structural components for the first F-35A Joint Strike Fighter when the aircraft's horizontal tails were installed at its Fort Worth, Texas, facility on 8 December

2005. Pratt & Whitney completed the first F135 engine at its Middletown, Connecticut, facility on 5 December 2005. Development of the F135, which is based on the F-22A's F119 engine, began in 1996, and P&W expects to achieve initial flight release (IFR) for the engine in the spring of 2006. Test engines have already logged more than 4,000 ground test hours in support of the F-35 System Development and Demonstration (SDD) phase. First flight of the initial F-35 is planned for fall 2006.

Raptor news

Lockheed Martin will build a Replacement Test Aircraft (RTA) for the F-22A programme and has received an $18 million contract from the USAF Aeronautical Systems Center covering long-lead effort for the RTA. The aircraft was included in the FY2006 defense appropriation and is funded as part of research development test and evaluation efforts.

An F-22A operated by the 411th Flight Test Squadron recently dropped its first 1,000-lb (454-kg) joint direct attack munition (JDAM) over the range at Edwards AFB, California. Although the Raptor has carried out seven supersonic JDAM separation flights since July, this mission marked the first where a guided weapon was deployed. Delivery of the weapon from high altitudes and supersonic speeds allows the Raptor to release the precision weapons at longer stand-off ranges than would otherwise be possible.

Global Hawk update

Northrop Grumman, the US Navy and USAF recently demonstrated the RQ-4A UAV's ability to detect airborne targets over the Pacific Ocean test ranges off southern California, as part of the Global Hawk Maritime Demonstration (GHMD).

Flown in support of exercise 'Trident Warrior 05', the 12.3-hour mission tested the radar's ability to locate and track airborne targets and maritime targets at ranges up to 200 km (108 nm). Sensor data was passed in real-time via the GHMD ground station and the Tactical Auxiliary Global Hawk System (TAGS), to USAF and Navy facilities on land and in the Atlantic Ocean aboard the USS *Iwo Jima* (LHD 7) and the USS *Mount Whitney* (LCC 20). The RQ-4A, one of two acquired by the navy, was operated by contractor personnel located at Edwards AFB, California. The air vehicles carried out four flights, accumulating 31.8 hours over a period of 12 days. Once the initial sensor testing has been completed at Edwards the two air vehicles will be flown to NAS Patuxent River, Maryland, where air test and evaluation squadron VX-20 will conduct flight tests and tactical experiments in support of the GHMD. The Navy RQ-4As differ from the USAF models in being equipped with new maritime search radar modes and a 360° LR100 signals intelligence sensor.

Raytheon recently delivered the first full production RQ-4 Global Hawk integrated sensor suite (ISS) to Northrop Grumman as part of the UAV's Lot 2 low-rate initial production (LRIP) programme. The configuration integrates synthetic aperture radar (SAR) and electro-optical/infrared (EO/IR) high-resolution imaging capability in a single system. Raytheon is currently building two basic ISS systems and one enhanced ISS as part of the programme's Lot 3 LRIP. The latter version extends the range capabilities of both the SAR and EO sensors by approximately 50 per cent.

The third development Global Hawk concluded a three-year deployment to Southwest Asia when it returned to Edwards AFB on 20 February. While deployed serial 98-2003 chalked up more than 4,800 flight hours in support of Operations 'Enduring Freedom', 'Iraqi Freedom' and 'Horn of Africa' flying 249 missions including 191 combat sorties.

Growler official

The Navy officially assigned the popular name 'Growler' to the EA-18G variant of the Super Hornet, which will replace the EA-6B Prowler in airborne electronic attack role. The nickname, which had been selected by personnel assigned to the electronic attack wing at NAS Whidbey Island, Washington, was already used informally to identify the aircraft. Based on the two-seat F/A-18F, the EA-18G will be equipped with an advanced version of the EA-6B's electronic attack suite including the AN/ALQ-218(V)2 Tactical Receiver, ALQ-99 Tactical Jamming System Pods and the Multi-Mission Advanced Tactical Terminal (MATT). It will also feature additional capabilities including a communications countermeasures system receiver. Two Super Hornets, referred to as EA-1 and EA-2, are being modified and the first is scheduled to fly in September 2006. The first four production Growlers were funded by the FY2006 defense appropriation.

Advanced Hawkeye progress

The US Navy recently completed a critical design review of Northrop Grumman's advanced E-2D Hawkeye airborne early warning aircraft and authorised the contractor to complete production of the two test aircraft. The aircraft are being built as part of a $2 bn system development and demonstration (SDD) contract from the Navy. The E-2D, which is the sixth generation of the Hawkeye, will combine battle-management, theatre air missile defense and multiple sensor-fusion capabilities into a single platform. The first E-2D will make its initial flight late in 2007, and the aircraft will achieve initial operational capability (IOC) in 2011. Current planning calls for the navy to purchase 75 examples.

The navy recently conducted tests flights of an E-2C airborne early warning aircraft that had been equipped with an inflight refueling probe. As part of the programme the Hawkeye demonstrated dry contacts with a KC-130F tanker and an F/A-18E equipped with a buddy refuelling store. Although additional testing is required the Navy hopes to retrofit its fleet of E-2C Hawkeye 2000s and early production E-2D Advanced Hawkeyes with the probe, but later E-2Ds would likely be equipped with probes during production.

C-130J update

Airmen from the USAF's 48th Airlift Squadron recently put the C-130J through a series of realistic tests at the Joint

MV-22B BuNo. 166491 is seen on a pre-delivery test flight over Texas. The first operational Block B aircraft – the 70th Osprey built – was handed over to the US Marine Corps on 8 December 2005.

Readiness Training Center (JRTC) in Fort Polk, Louisiana. The joint exercise, which demonstrated the aircraft's ability to resupply Army combat operations via engines-running offload (ERO) missions, air drops and at dirt landing strips, was evaluated by personnel from the USAF Operational Test and Evaluation Center (AFOTEC) at Edwards AFB, California. The operational evaluation moved to Eielson AFB, Alaska, in December for cold weather testing.

Although Pentagon officials earlier announced plans to terminate its five-year plan to purchase 60 C-130Js, the decision was subsequently reversed and more recently the Department of Defense announced plans to purchase 26 additional aircraft over the next two years at a cost of $1.15bn. Under the new plan the DoD will spend $430m in 2007 and $718m in 2008 to purchase 18 C-130J transports for the USAF and eight KC-130Js for the USMC. The two services currently operate 135 examples of the aircraft.

HC-130 replacement considered

USAF Special Operations Command (AFSOC) is considering replacing or updating the MC/HC-130 tankers used for refuelling combat search and rescue (CSAR) helicopters and CV-22B tiltrotor aircraft. The 36 MC/HC-130N/P tankers are on average 37 years old and, like many of the service's older C-130E transports, are beginning to show signs of fatigue. Accordingly, the service is preparing an initial capabilities document that will establish the requirements for a new Combat Rescue Tanker Replacement (CRT-X) programme. It has set a tentative date of 2011 for the CRT-X to achieve initial operational capability (IOC), and current planning calls for the procurement of 46 aircraft beginning in 2008. Besides aerial refuelling, the aircraft will be capable of delivering pararescue personnel via overt/covert airland or air drop delivery, and acting as airborne mission command (AMC) platforms. The service is also exploring the possibility of merging the CRT-X with its Joint Cargo Aircraft (JCA) project.

Already pursuing the USAF/US Army Joint Cargo Aircraft (JCA) project, Global Military Aircraft Systems (GMAS) has announced that it will propose the C-27J Spartan medium tactical airlifter to the USAF in response to the CRT-X requirement. GMAS is a joint venture between L-3 Communications and Alenia North America. The C-27J is also competing for

Canada's Fixed Wing Search and Rescue programme.

CSAR-X programme

The USAF Special Operations Command (AFSOC) formally unveiled its $8bn Combat Search and Rescue (CSAR-X) programme during October 2005 when it issued a request for proposals (RFP) associated with the 141-aircraft project. AFSOC intends to select a winner of the CSAR-X competition in August 2006 and the aircraft will achieve initial operational capability (IOC) in spring 2012. The initial order will include three non-developmental aircraft. Originally known as the Personnel Recovery Vehicle (PRV), the CSAR-X will replace the USAF's fleet of HH-60G combat rescue helicopters. Sikorsky Aircraft, Lockheed Martin and Boeing have also submitted proposals that are respectively based on the H-92, US101 and MH-47G helicopters.

Boeing recently demonstrated its MH-47G to the USAF at Nellis AFB, Nevada. The three-day demonstration included 20 hours of flight operations that highlighted the ability of the so-called HH-47 to conduct the CSAR mission.

Huey replacement plan

The USAF is defining the requirements for its Common Vertical Lift Support Program (CVLSP) helicopter. The CVLSP will replace the service's fleet of Bell UH-1N helicopters, which are primarily flown by four Air Force Space Command (AFSPC) Helicopter Squadrons (HS) in support of security forces assigned to missile wings. The UH-1Ns are also operated by Air Mobility Command's (AMC) 89th Airlift Wing (AW) and by the Pacfic Air Force's (PACAF) 374th AW in Japan for VIP duties. Air Education and Training Command's (AETC) 58th Special Operations Wing (SOW) and 336th Training Group (TRG) also use the aircraft

for training and training support duties. The service could purchase as many as 66 CVLSP helicopters, which were originally envisioned as a less complex version of the USAF's CSAR-X helicopter.

Tanker news

US government and military leaders recently announced their preference that the USAF acquire a fleet of Boeing 777s to fulfill the aerial refuelling role, citing the larger aircraft's ability to perform multiple missions. The announcement further muddies the USAF's attempt to acquire a fleet of 100 767 tankers. The Department of Defense pulled the plug on the USAF's plan to lease/purchase 100 B767 tankers at a cost of $17 bn in the wake of scandals in 2003. A new competition is expected, and Airbus and its partner Northrop Grumman plan to compete against the 767 by submitting a bid for its KC-30, which is based on the A330 airliner. Boeing is already building eight 767 tankers, comprising four each for the Japanese and Italian AFs.

A long awaited classified analysis of the USAF's tanker modernisation plans has determined that there is no immediate need to replace the service's KC-135 fleet. The analysis of alternatives carried out by the Rand Corporation reportedly recommends that the service look at budget constraints and other considerations before purchasing new aircraft. The study also recommends adapting medium-to-large commercial aircraft for the mission, and narrowed the field to six airframes including the Boeing 767, 777, 787 and 747, and the Airbus A330 and A340. The USAF is anxious to begin replacing the KC-135 fleet, which is more than 40 years old, and expects to start a formal competition later this year. Its proposed budget for 2007 includes $250 million that is earmarked for the programme. The service hopes to have the new aircraft in service around 2011/2012.

UPGRADES AND MODIFICATIONS

Australia

Chinook upgrades

Australia has announced plans to provide the Army's fleet of Chinook helicopters with a $25 million upgrade that will equip the aircraft with electronic warfare and ballistic protection, and advanced communications equipment. Additionally, the aircraft will be equipped to carry out aeromedical evacuation missions and will receive improved gun mounts. The 5th

Regiment's C Squadron operates six CH-47Ds from Townsville, Queensland.

Brazil

Orion flies

The first of 12 P-3As destined for the Fuerza Aérea Brasileira made its first flight at Davis Monthan AFB, Arizona, after being removed from storage at the Aerospace Maintenance and Regeneration Center (AMARC) on 20 December 2005.

High over Chesapeake Bay near Patuxent River, an E-2C from the Naval Air Warfare Center demonstrates its ability to refuel from a KC-130F. Both aircraft are operated by air test and evaluation squadron VX-20.

On 5 November 2005 US Helicopter, a division of Bell Aerospace Services Inc., rolled out the first of 24 TH-1H helicopters for the USAF at Randolph AFB, Texas. The TH-1Hs are based at Fort Rucker, Ozark, Alabama, and are used for rotary-wing training. The TH-1H is a UH-1H reworked for the training role with the Huey II upgrade kit.

The aircraft was subsequently flown to Chico AP, California, where Aero Union prepared it for a ferry flight to Madrid, Spain. The former BuNo. 152180, now serialled FAB7200, arrived in Spain on 11 January. EADS/CASA will overhaul and modernise the aircraft, equipping it with new systems that include search radar, anti-submarine warfare, infrared/electro-optical, and acoustic sensors, magnetic anomaly detector and a datalink. Six operator consoles will be equipped with the fully integrated tactical system (FITS). Only eight Orions will be modernised with the remainder being retained for spares. The initial modernised P-3AM will be delivered to Brazil in 2008.

Bulgaria

Helicopter enhancement moves
Selection was announced in December by the Defence Ministry in Sofia of Elbit Systems Ltd and Lockheed Martin as preferred bidders to upgrade 12 Mi-24 attack and six Mi-17 transport helicopters to NATO standards. Subject to completion of negotiations and contract signature, Elbit Systems will be prime project-contractor. The reportedly $57m programme will involve installation of new digital avionics and missions systems for enhanced combat capabilities and full NATO interoperability, following Bulgaria's entry into the alliance in March 2004.

On 28 January Bulgarian Defence Minister Svinarov and Eurocopter Senior VP Sales & Marketing Luc Barrière signed a contract in Sofia for procurement of 12 AS 532AL Cougar and six AS 565MB Panther twin-turboshaft helicopters. The 12 Cougars for the Bulgarian air force (BVVS) will comprise eight for tactical transport duties, the first three being delivered by late 2006, and four equipped for Combat SAR missions.

The six Panthers for Bulgarian Naval Aviation (ANB) will undertake sea surveillance, ASW/ASuW and SAR missions. All 18 helicopters are due for delivery by 2008.

Canada

Aircraft upgrades progress
On 6 January the Canadian Forces reached half-way completion of their Phase I CF-18A/B combat aircraft modernisation, from Boeing International's delivery of the 40th upgraded Hornet. The company's $880m 2001 contract includes procurement and installation of an advanced upgrade package to extend CF-18 operating lives until at least 2017. Based on the US Navy's F/A-18 upgrade programme, the package includes a new radar, Have-Quick jam-resistant radios, mission-computers, combined IFF interrogator/transponder, stores-management systems, and embedded GPS/INS.

Due for completion by summer 2006, the aircraft modifications are being installed by L-3 MAS at its Mirabel, Quebec, facilities. Phase I is the larger of the two-part programme, for completion in parallel with several other upgrades, from new simulators to advanced AAMs, already operationally proven by the US Navy.

Last November, Lockheed Martin received a $C14.5m ($12m) contract for integration and support of a new electro-optical infra-red (EO/IR) surveillance capability for Canada's Lockheed CP-140 (P-3) Aurora maritime-patrol aircraft. As prime-contractor, Lockheed Martin Canada (LMC) will integrate and install L-3 WESCAM MX-20 long-range surveillance cameras on board five Auroras, with options for 10 more.

Chile

Fighter update
Lockheed Martin has received a $7.5 million contract that covers the modification of 18 Dutch F-16A/B aircraft to a modified mid-life update (MLU) configuration for Chile. In support of this effort, the aircraft's capability to deliver precision weapons and AGM-88 HARM missiles will be removed. Fokker will overhaul the fighters at its Woensdrecht facility in the Netherlands prior to their delivery to Chile in 2006 and 2007.

Indonesia

C-130s returned to service
US aid to Indonesia during the tsunami relief operations over the turn of the year included assistance in restoring more than a dozen of the TNI-AU's 25 Lockheed C-130s to airworthy condition. Only nine of the C-130s – mostly C-130H/H-30s, apart from nine elderly C-130Bs – had been maintained in service, following a long-term US ban on arms and military spares sales to Indonesia. Two technicians from Lockheed Martin's Air Logistics Center in Greenville, S.C., arrived on 16 January with spare parts to begin work on five C-130Hs at Jakarta's military airport. They were assisted by crews from the USAF's 517th Airlift Squadron at Elmendorf Air Force Base, Alaska, who arrived on 6 January with four C-130s and 120 personnel. Working with Indonesian personnel, they restored the first five TNI-AU C-130s to flight status and immediate service within a week or so of their arrival.

Iraq

Huey updates
The 16 UH-1H helicopters donated to Iraq by the Royal Jordanian Air Force will be upgraded to Huey II configuration by ARINC Engineering and US Helicopter at the latter's facility in Ozark, Alabama. As part of the modifications the helicopters will be equipped with more powerful engines, transmissions and gearboxes, plus new tail booms, main and tail rotors. The first UH-1H will arrive in Alabama for the 6- to 7-month upgrade by March, and the upgraded Huey II is expected to be returned in late 2006.

New Zealand

C-130H upgrade contract award
L-3 Communications announced in December a long-awaited $NZ226m ($161.4m) contract award to its Spar Aerospace Ltd (L-3 Spar) Canadian subsidiary for a 15-year life-extension programme for the Royal New Zealand air force's five Lockheed C-130Hs. This includes replacement of mechanical, avionics and structural components, plus design and installation of flight deck communications and navigation improvements. Modifications will begin in early 2006 on the first aircraft at L-3 Spar's facility in Edmonton, Canada, followed by the second to fifth aircraft at Safe Air Ltd's Blenheim facility in New Zealand.

The new contract immediately followed NZ Defence Ministry selection in December of L-3's Integrated Systems subsidiary, to upgrade mission and nav/comms systems for the RNZAF's six Lockheed P-3K maritime-patrol/SAR aircraft. This involves supply by L-3 Communications WESCAM subsidiary of electro-optical and infra-red (EO/IR) imaging sensors for the RNZAF's P-3K Project Guardian upgrade programme. RNZAF P-3Ks, already incorporating structural and some systems upgrades, will receive new mission-management systems and WESCAM MX-20 EO/IR imaging-turrets, plus ground-based support and training, from L-3 Communications' Integrated Systems Division from early 2005.

Two RNZAF Boeing B757-200s operated as passenger transports since replacing two Boeing 727-100s in early May 2003 are being modified and upgraded for additional cargo roles, to FAA Supplementary Type Certificate (STC) standards. Last July, the NZ government nominated Singapore Technologies Aerospace (ST Aero) Mobile Aerospace Engineering (MAE) as prime-contractor for the modifications, at MAE's Alabama facilities.

Changes costing NZ$100-$200m ($140-279m) will include installation of a cargo door, RB211-535-E4B engine thrust increase, and upgraded civil and military communication, navigation, surveillance/air traffic management (CNS/ATM) capabilities, from late 2005.

Peru

Service news
Peruvian government signature was expected late last year with Russia's Rosoboronexport arms sales agency for a contract worth about $35m to return a number of FAP support aircraft, grounded from spares shortages, to operational service. Some eight FAP Antonov An-32 twin-turboprop tactical transports and 18 Mil Mi-17 helicopters were reportedly involved in this programme, from large numbers of these and other types from earlier Soviet procurement.

Plans to upgrade 17 remaining FAP MiG-29S/SE 'Fulcrum-Cs', two two-seat MiG-29UB combat trainers, and 10/8 Sukhoi Su-25/UBs bought from Belarus in 1996-98, and postponed last year through lack of spares and technical support, may also be revived, if the necessary funding can be found. Delivered from 1976, the survivors of the FAP's 32/16 Sukhoi Su-20/-22M2K 'Fitter-F/J' strike/interceptors and four Su-22UM3 'Fitter-G' two-seat combat trainers are overdue for further upgrades, after interim 1984 modernisation by SEMAN-Peru. One of the Su-22UM3s crashed on 17 December south of Lima, killing both crew members.

Spain

First production EF-18 upgrade
Formal delivery took place at EADS Military Aircraft facilities at Getafe on 2 February of the first EdA Boeing EF-18 Hornet to have received a production mid-life upgrades (MLU) to its avionics and missions systems, from a Euro186m ($241m) December 2003 contract. Some 65 single-seat EF-18As and two two-seat EF-18Bs, operated by the 12th and 15th Wings at Torrejón and Zaragoza Air Bases respectively, are being fitted by EADS CASA with new digital missions systems from several suppliers, including Kaiser

Electronics, Ericsson, Indra, and SAAB Avionics.

The software team of the EdA's Armament and Experimental Logistics Centre (CLAEX) and EADS CASA Military Aircraft have actively collaborated in the design and development of the integration and testing of the new avionics equipment. This includes new tactical computers, high speed multi-processors, two multi-function colour display screens, INS/GPS, Link 16 Have-Quick II secure radios, IFF, night-vision goggle capability, and complete development of the modified system integration software.

United Kingdom

Tornado F.3 upgrades continue

Despite firm orders and initial deliveries of Eurofighter Typhoons to begin their replacement, further upgrades and technical support for the RAF's 130 or so Panavia Tornado F.Mk 3 air defence fighters are planned from a £25m ($46.85m) December MoD Smart Acquisition contract with BAE Systems. The new F.3 Sustainment Programme (FSP) will integrate the latest air defence weapons, comprising Raytheon's AIM-120C-5 advanced medium-range air-to-air missile (AMRAAM), and MBDA's advanced short-range ASRAAM FOC2 into the Tornado's mission system.

Also included are radar and on-board computer software changes that significantly enhance missile targeting through improved onboard processing and display systems. Tornado Integrated Project Team Leader Air Commodore Nigel Bairsto said that the contract should deliver economies of over £14m ($26.24m) through rapid

prototyping, a combined trials team, and lean engineering. As prime contractor, in partnership with QinetiQ, AMS, BAE Systems Avionics, Raytheon, and MBDA, BAE Systems will oversee Tornado F.3 vendor management, assets-integration, and ensure base-level FSP embodiment capabilities.

United States

New AWACS model

The USAF has allotted the mission design series (MDS) designation E-3G to 32 Block 30 E-3B and Block 35 E-3C airborne early warning aircraft that will be upgraded to a new common Block 40 configuration. The USAF plans to equip its entire fleet of Sentry's with the modifications, which will provide the aircraft with new computers, an open architecture, increased data integration, and network-centric capabilities. The first modernised E-3G is scheduled to enter service by 2009.

Stratofortress avionics update

The USAF's Oklahoma City Air Logistics Center has awarded Lockheed Martin a $28 million contract associated with the B-52 Avionics Midlife Improvement (AMI) Program. The AMI will upgrade the B-52H's offensive avionics system and includes the replacement of the inertial navigation system (INS), the avionics control unit, the data transfer system and associated hardware and software. Testing of the system began in December 2003 at Edwards AFB, California.

C-12J cockpit upgrade

Vertex Aerospace, a component of L3

A pair of F-15As from the Hawaii Air National Guard's 199th Fighter Squadron escorts C-17A 05-5146 to its new home at Hickam AFB. By the end of 2006 the Hawaiian base will have eight C-17As, jointly operated by the active-duty 15th Airlift Wing and the ANG's 154th Airlift Wing.

Communications Integrated Systems, has received a $7.3 million contract from the Oklahoma City Air Logistics Center under which it will equip six C-12Js operated by the USAF and US Army with EFIS modifications.

Dolphin re-engine programme

American Eurocopter has received a contract from the Integrated Coast Guard Systems (ICGS) team under which it will re-engine and upgrade 11 US Coast Guard

HH-65B Dolphin helicopter to HH-65C configuration. The effort, which will be completed in late 2006, will be carried out at the contractor's Columbus, Mississippi, facility and includes an option for six additional conversions. The Coast Guard is currently carrying out HH-65B to C conversions at its Aircraft and Supply Center in Elizabeth City, North Carolina. Opening the second production line will accelerate the rate at which the upgraded helicopters can be returned to service.

PROCUREMENT AND DELIVERIES

Afghanistan

Deliveries

The Russian Defense Ministry has provided Afghanistan with $30 million in military equipment that includes a pair of Mi-24 helicopters and spare parts for An-26 and An-32 transport aircraft. Two additional Mi-24s will be delivered, along with a pair of L-39 jet trainers, An-26 and An-32 airlift aircraft and six spare helicopter engines.

Algeria

EC 225 VIP helicopter delivered

On 22 December Eurocopter officially delivered the first series production EC 225 Super Puma/Cougar executive helicopter to roll off the Marignane assembly lines to the Ministerial Air Liaison Group (GLAM) of the Algerian Republic. This twin-Turbomeca Makila 2A turboshaft 11-tonne helicopter made its initial flight in November 2000, and has already been chosen by 32 heads of state or governments.

In July 2004, the EC 225 demonstrated its embodiment of the most recent JAR 29 regulations, and was awarded its IFR certification by the new European Aviation Safety Agency airworthiness authority (EASA). The latest VIP example will join the AQAJAJ's prestigious transport squadron, which serves the Algerian Presidency.

Burkina Faso

Helicopter delivery

Russia's Rosoboronexport defence export company delivered a pair of Mi-35 Hind combat transport helicopters to Burkina Faso in western Africa during December 2005. Manufactured at Rostvertol, the helicopters were airlifted to Ouagadougou, by an An-22, where they were reassembled and flight tested.

Canada

UAV order

Canada's Department of National Defence (DND) has announced the purchase of five new SAGEM Sperwer UAVs for use by its units deployed to Afghanistan. The Army had previously operated the French built UAV in Afghanistan during 2003/04 when it purchased six UAVs and associated equipment at a cost of $33 million. During its initial deployment the UAVs were launched 36 times but suffered 14 incidents including five accidents.

Tactical airlift plans

The DND announced that it is proceeding with a competitive procurement of a new fleet of tactical airlift aircraft for the air force. The plan calls for the purchase of 16 new aircraft at a cost of $3.5-4.3 billion including 20 years worth of support. The new aircraft will replace 13 CC-130E airlifters and the DND expects to select a

winner in 2007 and take delivery of the aircraft between 2010 and 2012.

Colombia

Tucanos ordered

The Colombian government has ordered 25 Super Tucano training/light attack aircraft from Embraer at a cost of $235 million. The Fuerza Aérea Colombiana, which already operates 14 older examples of the Brazilian aircraft, will use the Tucanos for internal security operations and border patrol. Embraer will deliver six examples in 2006, 10 in 2007 and the remainder in 2008. The FAC also has a requirement for SIGINT aircraft, similar to Brazil's specially-equipped EMBRAER EMB-145s.

Recent FAC upgrades have included a seventh Douglas AC-47 Dakota gunship,

with a new fire-control system, FLIR and NVGs for night-attack roles. Five of these have also undergone Basler PT6A-turboprop conversions, as AC-47T Fantasmas (Phantoms), from a $6.4m US aid programme. This has totalled $3.7bn since 1997, compared with Colombia's 2004 defence budget of less than $3bn.

Denmark

EH101 deliveries begin

Deliveries are now in progress of the 14 EH101s being produced at AgustaWestland's Yeovil factory for the RDAF, officially named Merlin Joint Supporters at an official ceremony late last year by Major General Klaus Axelsen of Denmark's Air Materiel Command. The new name reflects the EH101's multi-role

The third and fourth F-15Ks for the Republic of Korea Air Force passed through Hickam AB in Hawaii in December 2005. Note the slightly curved contours of the nozzles for the General Electric F110 engines, the F-15K being the first production Eagle version to have this powerplant.

Above: Piloted by Lt Col James Hecker, the 27th Fighter Squadron commander, an F-22A takes off on 21 January 2006 from Langley AFB to fly the Raptor's first operational sortie, an Operation Noble Eagle homeland defence mission.

Below: On 3 March the USAF's second operational F-22A squadron – the 94th Fighter Squadron – received its first two aircraft. Here 04-062 touches down at Langley to join the 'Hat in the Ring' squadron.

capabilities in supporting all three Danish services, in addition to its basic SAR and tactical transport roles. Funding is now reportedly being sought for the RDAF to take up its options on another four EH101s.

Egypt

More C-130s
Having taken delivery of 23 Lockheed C-130Hs and three stretched C-130H-30s by 1978 and 1990, the Egyptian air force (EAF) is receiving another three C-130Hs and related spares from a new $30.69m US Foreign Military Sales fixed-price contract announced on 10 December 2004. The aircraft are being refurbished by Lockheed Martin Marietta, for imminent delivery.

Transport purchase
The Egyptian Air Force recently purchased a single An-74TK-200A, from the Kharkov State Aircraft Manufacturing Company (KSAMC) in the Ukraine, at a cost of $34 million. The aircraft entered service in September 2005 and Egypt plans to purchase as many as 18 additional examples, which will feature advanced cockpit displays and electronics.

Equatorial Guinea

Transport plans
The west African nation is considering the purchase of An-74 transport aircraft from the Kharkov State Aircraft Manufacturing Company (KSAMC) in the Ukraine. It will initially purchase one aircraft, which will be tested under operational conditions. The country is considering the An-74 for use in the VIP transport, passenger, cargo and cargo-passenger combination configurations.

France

First EC 725s delivered
Deliveries started earlier this year to the French air force (AdA) of the first of six Eurocopter EC 725 Cougar Mk II+ twin-turboshaft helicopters equipped for combat search and rescue (CSAR). Another eight EC 725s are also being delivered by 2006 for operation by French special forces.

Hungary

Initial Gripen roll-out
Formal roll-out took place at SAAB's Linköping factory in Sweden on 25 January of the first of 14 Hungarian JAS 39EBS-HU Gripens to be produced for the 15-year lease-to-buy contract signed in early 2003. Production is also well advanced of another five Hungarian Gripens, for delivery from March 2006. After final assembly, the Gripens will undergo ground and flight acceptance tests by SAAB and official FMV pilots, although arrival of the first five Hungarian LeRP pilots in Sweden was reported in January, to start their year-long JAS 39 conversion training with the SAF's F7 (Skaraborg) Wing at Såtenäs.

After graduation, they will return to Hungary as Gripen instructor pilots, although initial LeRP conversions will start in Sweden in mid-2006. First Hungarian air force (LeRP) unit to operate the Gripen will be the 59th Fighter Wing, at Kecskemét, from March 2006.

AMRAAM confirmed
A US Letter of Offer and Acceptance (LOA) valued at $25,389,904, for acquisition of 40 Raytheon AIM-120C-5 AMRAAMs and associated equipment, was signed in mid-December by the Hungarian MoD. The contract is conditional on successful airframe/AMRAAM integration certification with Hungary's new JAS 39C/D Gripens.

Work on AMRAAM/Gripen integration is already underway in Sweden, from a separate deal between Raytheon and SAAB/BAE, plus a Swedish Materiel Administration (FMV) order last March for two AIM-120C-5s. The Gripen is currently certified only for the earlier AIM-120B AMRAAM, so more integration, related mainly to software, is required, prior to planned 2006-07 weapons delivery to Hungary. This will allow the LeRP to meet its NATO Quick Reaction Alert (QRA) tasks from January 2009.

The AMRAAM offer was part of a larger US Gripen package comprising stockpiled AIM-9M Sidewinder AAMs, newly manufactured GBU-10, -12, -16 laser guidance kits for Mk 82, 83, 84 bombs, and Raytheon AGM-65H/K Maverick ASMs. These were all rejected by Hungary, however, on cost grounds. Hungarian MoD tenders were expected to be invited earlier this year, to fulfil LeRP Gripen requirements for short range AAMs and air-to-ground weapons, for mid-2005 selection. The LeRP favours the Bodenseewerk Geratetechnik/SAAB Dynamics IRIS-T with a helmet-cued sight, but funding considerations may limit Hungarian short-range AAM procurement to stockpiled German or Swedish AIM-9L Sidewinders.

India

MiG-29K contract details
In a recent written parliamentary reply, Defence Minister S.P. Mukherjee confirmed details of the contract signed on 20 January 2004 with Rosoboronexport for the purchase of 16 MiG 29K carrier-based combat aircraft for the Indian Navy. Costing $740.35m, their delivery was due to start in June 2007, to provide integral air capability to the Navy at extended ranges. Offensive air strikes against shore targets would be in direct support of land operations by the Army.

Su-30 programme progress
Recent major milestones in India's Sukhoi Su-30K/MKI procurement and production programme included formal roll-out on 28 November of the first of 140 Su-30MKIs being built by Hindustan Aeronautics Ltd (HAL) at its Ozar air base facility at Nasik. This preceded delivery of the 50th and last Su-30 to various standards from Irkut production, comprising 38 Su-30Ks, including 12 with canard foreplanes, and the first dozen full-standard Su-30MKIs with added thrust-vectoring and advanced mission system avionics, on 26 December.

OAO Irkut Scientific and Production Corporation (NPK Irkut) president Aleksei Fedorov then announced that Su-30 programme contracts signed between India and Russia totalled nearly $5bn. He said that the first Indian contract, signed in November 1996, for an initial 40 Irkutsk Aviation Plant-built Su-30s, and worth nearly $1.5bn, was extended in 1998 for another 10 Su-30s. All the earlier IAF Su-30Ks are scheduled for HAL upgrades to full Su-30MKI standards.

Interestingly, he referred to HAL's Su-30MKI licensed production programme, officially quoted in Delhi as costing Rs22,122.78 crore, or $4.8bn, as involving 140 aircraft, and not the reduc-tion to 120 quoted in some recent reports. Rosoboronexport is also quoting 140 Su-30MKIs as HAL's production target, for completion by 2017. Irkut and HAL have also agreed to manufacture some Su-30 components in India for export customers.

Irkut listed Su-30MKI avionics as including the most advanced developments of companies from Russia, India, France, Israel and the UK, integrated by the Ramenskoye Design Bureau (OKB). The Su-30MKI's NIIP Tikhomirov N011 Bars (Panther) radar has a phased-array antenna, allowing simultaneous tracking and attack of several air and surface targets. HAL also license-produces some avionics, together with the Su-30MKI's thrust-vectored Saturn/UMPO AL-31FP turbofans – the first to enter world operational service.

OAO Saturn Scientific Production Association (NPO Saturn) General Director Yuri Lastochkin, has quoted new service life figures for IAF Su-30MKI AL-31FP engines and their rotating nozzle assemblies. These comprise 1,000 and 500 flying-hours, respectively, between major overhauls. Initial IAF Su-30MKI experience has reportedly indicated some difficulties in reaching even these modest targets, and overall AL-31FP life is currently only 2,000 hours.

The Su-30s have been among the main users of the IAF's Ilyushin Il-78MK tanker/transports, of which the sixth and last was delivered from the Tashkent Chkalov Aviation Production Association (TAPOiCh) in December, from a late-2001 $152m contract. The IAF received its first Il-78 in March 2003, and air-refuelling has become a standard procedure to extend its operations.

Navy modernisation
Northrop Grumman recently held discussions with the Indian Navy regarding the sale of six to eight E-2C Hawkeye 2000s. The service recently had a first-hand look at Hawkeye operations during Exercise Malabar 05 when an Indian Ka-31 controller flew aboard a VAW-117 E-2C operating from the aircraft-carrier USS *Nimitz* (CVN 68). India would likely operate the Hawkeyes from land bases rather than its new aircraft-carrier.

The Indian Navy is also considering a British offer to provide six to eight retired Sea Harrier for use in the training role.

Ireland

More helicopter orders
After a competitive evaluation programme from mid-2004 tenders, Irish Defence Minister Willie O'Dea signed a Euro49m ($66.57m) contract with AgustaWestland in January, for acquisition of four new Air Corps AB139 utility helicopters. Delivery of the first two from Bell Agusta Aerospace is scheduled in 2006, followed by the two others in 2007, for IAC operational, VIP transport, air ambulance, and training roles. The IAC is also acquiring two more EC135 light utility helicopters for general support roles, from Eurocopter SAS, through an additional Euro11m ($14.94m) contract.

Japan

Apache Longbow delivered
Boeing formally delivered the first AH-64D Apache Longbow helicopters

On 17 February 2006 the second Gulfstream G500 Nachshon Sigint aircraft (Shavit, comet) arrived in Israel. Shortly after, on 26 February, the Israel Air and Space Force inaugurated a Shavit Flight at Lod air base. A third Sigint aircraft is expected in 2006, along with the first AEW platform.

destined for Japan to Fuji Heavy Industries (FHI) at its Mesa, Arizona, facility on 15 December 2005. Designated the AH-64DJP, Japan's Apaches are the first production examples capable of operating the air-to-air Stinger missile. They will be delivered to the Japanese government by FHI in March 2006. Although Boeing produced the initial pair of AH-64DJPs, FHI is assembling the subsequent airframes, for the Japan Ground Self Defense Force, locally in Japan under a license from Boeing.

Coast Guard purchases
As well as buying two more Saab 340SAR-200s, the Coast Guard has followed the lead set by the Ground Self Defense Force (JGSDF) and ordered a pair of EC 225 helicopters, which will be used for both transport and SAR duties. The air arm, which is a component of the Ministry of Land, Infrastructure and Transport, already operates earlier models of the AS332 Super Puma.

Jordan

Viper deal
The Jordanian and Netherlands governments have reached an agreement under which the former will purchase three F-16B(MLU) aircraft for use by the Royal Jordanian Air Force. The fighters will be delivered in October 2006.

Libya

New Antonov
The Libyan Air Force has ordered a new An-74TK-300VIP from the Kharkov State Aircraft Manufacturing Company in the Ukraine. The aircraft will join two An-74TK-200C flying ambulances that are already in service.

Malaysia

Work starts on Su-30MKMs
According to an ITAR-TASS news report, work has begun at the Irkutsk Aviation Production Association (IAPO) – a subsidiary of the Irkut Scientific and Production Corporation – on the fulfillment of a contract for the delivery to Malaysia of 18 Su-30MKM fighters. Malaysian air force specialists are at the Irkyutsk factory to oversee the purchase and installation of customer-specific equipment. A first delivery is in late 2006.

Airlifters selected
The Malaysian government joined the Airbus A400M programme recently when it agreed to purchase four of the military airlifters. Malaysian industries will receive offset work as part of the package. The first A400M is scheduled to fly in 2007.

Nepal

ALHs ordered
Already operating 11 helicopters of five different types, including five Mil Mi-17s

delivered in 2000, the Royal Nepalese Army Air Service is receiving two multi-role HAL Dhruv advanced light helicopters (ALHs), with a reported unit cost of Rs250m ($5.76m), to augment its high-altitude fleet. Powered by two 801-kW (1,074-shp) Turboméca TM333-2B turboshafts, some 35 ALHs have so far been delivered to Indian armed forces and government agencies since mid-2002. About half the 120 Dhruvs required by Indian army aviation will incorporate integrated weapons systems for attack missions, with the remainder serving in utility and transport roles.

The 13-seat Dhruv, which incorporates about 60 per cenmt foreign components, has been demonstrated and evaluated as far afield as Chile, where President Ricardo Lagos recently expressed interest in possibly acquiring four or more. A single ALH was also delivered to the Israeli air force (IDF/AF) last December, and HAL production is building up towards one per week. HAL and Turboméca are also co-developing a higher-powered Shakti version of the TM333 for improving the Dhruv's hot and high-altitude performance.

Oman

RTM322s to power NH90s
The Royal Air Force of Oman (RAFO) has selected the Rolls-Royce Turboméca (RRTM) RTM322 turboshaft to power its new fleet of 20 Tactical Troop Lift (TTH) variants of the twin-engine NH90 multi-role helicopter. RAFO thus becomes the ninth NH90 customer to order the RTM322, after Finland, France, Germany, Greece, Netherlands, Norway, Portugal and Sweden, and will benefit from its upgraded 1 937-kW (2,600-shp) version.

Selected by 90 per cent of NH90 customers to date, the RTM322 has flown in all five prototypes, and also powers the first production aircraft. RTM322 module and engine assembly lines have been established in Finland, France, Germany, Norway and the UK, for the broad NH90 customer base. The RTM322 has also been selected for about 75 per cent of EH101 orders, the latest being from the Japanese Defence Agency.

Pakistan

F-16 provision deferred
Late 2004 press reports quoting PAF Chief of Air Staff, Air Chief Marshal Kaleem Saadat, as claiming positive indications from the US government concerning fulfilment of Pakistan's requirements for more Lockheed Martin F-16s, have still to be borne out. In return for Pakistani military assistance in anti-terrorist operations along the Afghan borders, Pakistan was declared a non-NATO US ally in mid-2004, and shortly before an early February meeting of the joint US-led Defence Consultative Group, ACM Saadat said that 18 Raytheon AMRAAM-armed F-16s were being

offered to the PAF, pending Congressional approval.

These were required to supplement some 30 survivors from 28 F-16As and 12 two-seat F-16Bs delivered from 1983, following the 1990 US embargo on 13/15 F-16A/B OCU versions costing $658 million, which were built by Lockheed Martin and paid for by Pakistan, but never delivered. The Islamabad government received $2.42bn in US military aid between January 2002 and September 2004, which included six surplus Lockheed C-130 transports worth $75m to improve PAF capabilities in defence and humanitarian missions.

Later in February, a Washington statement said that the US was still considering F-16 sales to Pakistan, which would be dealt with at an "appropriate" time. In view of India's potential interest, however, it seems unlikely that the US could release F-16s to the IAF while vetoing them for Pakistan.

Herc arrives
The first of six 'new' C-130E transports destined for the Pakistan Air Force arrived in Islamabad in early November carrying relief supplies for victims of the country's recent earthquake. Lockheed Martin is delivering the aircraft and upgrading the avionics of the PAF's existing Hercules fleet as part of a $75 million contract. Five of the C-130Es previously belonged to the Royal Australian Air Force and were taken in trade by Lockheed Martin as part of a deal that sent new C-130Js to Australia. The sixth operational aircraft was reportedly operated by the Argentine Air Force.

Portugal

EH101 deliveries begin
Formal delivery took place in December at AgustaWestland's Vergiate facility of the first of 12 AgustaWestland EH101 Merlin helicopters ordered by the Portuguese government for combat SAR and fishery protection. The FAP is expected to receive all its EH101s by the year-end. As part of the Portuguese contract, AgustaWestland has established partnerships with Portuguese industry, in aerospace and other sectors, to satisfy Portuguese government

industrial compensation and collaboration requirements.

Saudi Arabia

Modernisation plan announced
Saudi Arabia's defence minister has confirmed that the country intends to purchase and upgrade up to 200 aircraft as part of a deal that could be worth as much as $68.8 billion over the next 20 years. Although the details of the plan will not be released until March 2006, it reportedly includes 48-72 new Eurofighter Typhoons and 60 Hawk trainers, and upgrades for 80 Tornado strike aircraft. In order to facilitate an early delivery the UK Ministry of Defence has already approved the release of 24 Tranche 2 Typhoons, destined for the Royal Air Force, which will be diverted to the Royal Saudi Air Force beginning in 2008.

Singapore

Seahawks for naval air arm
Formation of a naval aviation element of the Singapore armed forces was signalled in January, from MINDEF contract signature with Sikorsky to acquire six new S-70B Seahawk helicopters. Equipped with advanced anti-surface/submarine warfare sensors and weapons, the S-70Bs will operate from the Republic of Singapore Navy's (RSN) new frigates, following scheduled Seahawk deliveries between 2008-10.

Training helicopters ordered
Singapore Technologies Aerospace (ST Aerospace) will provide the Republic of Singapore Air Force with new training helicopters under a $120 million contract awarded by the Ministry of Defence. ST Aerospace will purchase a fleet of five Eurocopter EC120 helicopters that will be operated by the RSAF for a period of 20 years. The contractor, which will maintain and support the aircraft, already support the RSAF transport wing training requirements using King Air C90 aircraft.

Strike Eagle contract award
Singapore and Boeing have agreed to the terms of a contract that includes the production of 12 F-15SG and an option

The first VH-71 Presidential helicopter test vehicle lands at NAS Patuxent River on 2 November 2005. Test vehicle one (TV1, MM81495) is being used to train pilots and familiarise maintainers with the new platform. The new Presidential helicopter hangar and support facility can be seen in the background.

This is one of two Saab 340SAR-200 aircraft that are operated by the Japan Coast Guard from Kansai Airport near Osaka. Two more aircraft have been ordered. Based on secondhand Saab 340B Plus aircraft, the SAR-200 version has a 360° search radar, enlarged observation windows, FLIR, and emergency liferaft-dropping capability.

for eight additional aircraft. The first F-15SG, which is an advanced version of the USAF F-15E will be delivered in 2008. Although similar to the Republic of Korea's F-15K (and also powered by the General Electrics F110), the fighters will feature the AN/APG-63(V)3 active electronically scanned array (AESA) radar.

Spain

More VIP helicopters
The EdA's government transport Escuadon 402, based at Madrid's Cuatro Vientos air base, was reinforced last December by deliveries from Eurocopter España of two AS 532UL Cougars. The two VIP helicopters, which were ordered in June 2003, will join the existing EdA fleet of four Eurocopter AS 332M1s and two AS 332B Super Pumas, for transport of senior officials. The AS 532Uls are receiving full Spanish civil airworthiness type certification, and their Euro33m ($42.75m) contract also includes support services, training and spare parts. Eurocopter España will undertake personnel training and logistical support, as well as the industrial activities included in the contract.

Sudan

An-74 sales
The Kharkov State Aircraft Manufacturing Company in the Ukraine recently sold four An-74TK-200 and two An-74TK-300 transports to Sudan at a cost of $102 million.

Turkey

ATR 72 nominated for MR/ASW
Selection by Turkey's Undersecreteriat for Defence Industries for Turkish Navy Aviation (TCBH) procurement of 10 twin-PWC PW127E-turboprop ATR 72MP-500 maritime patrol and anti-submarine surveil-

lance aircraft, was announced by Alenia Aeronautica in January. Contract negotiations were then started for the cost-competitive medium-range ATR-72s, which will be equipped with comprehensive Thales mission systems.

These have been selected by several countries, including France, Japan and Turkey. Designed for maximum flexibility, Thales Airborne Maritime Situation Control System (AMASCOS) meets all current and future operational requirements in search-and-rescue, EEZ surveillance, anti-surface and ASW roles. With its modular design, AMASCOS is designed around an advanced tactical command subsystem, and accommodates a broad range of sensor types including the Ocean Master Mk II radar, ESM, acoustic and communication systems.

AMASCOS is part of the $400m September 2002 MELTEM contract, for which Thales is supplying 19 systems to the Turkish Navy and Coast Guard, for installation in nine previously-ordered Airtech CN-235Ms and 10 new-build aircraft, from TAI production.

United Kingdom

Hawk 128 contract signed
In December, the Ministry of Defence (MoD) announced the award to BAE Systems of an initial Design and Development Contract (DDC), worth some £158.5m ($298m), for the RAF's new Hawk 128 advanced jet-trainer (AJT). The MoD will work closely with BAE Systems to manage the design of the digital avionics architecture, introduce a modern glass cockpit environment, and deliver two trials aircraft from BAE's Brough factory in Yorkshire, to support the development and test flying programmes.

Announced on 30 July 2003, the programme will involve RAF receipt of 20

Hawk Mk 128 trainers, with an option for a further 24, for pilots selected to fly the Tornado GR.Mk 4, Harrier GR.Mk 7/9, Typhoon and later, the F-35 Joint Strike Fighter. The total AJT programme is worth £3.5bn ($6.58bn), which includes the £800m ($150m) acquisition cost and the cost of operating, supporting and maintaining the aircraft over 25 years. Exact aircraft numbers, delivery schedule and In-Service Date will all be set at the time of the main investment decision, currently planned for early 2006.

United States

Trainers ordered
Raytheon Aircraft has received a $268.9 million contract from the Air Force Materiel Command covering the production of 64 Lot 13 T-6A training aircraft.

E-10 contract award
The USAF Electronic Systems Center has awarded Northrop Grumman a $280 million contract associated with the E-10A Battle Management Command and Control (BMC²) aircraft. The next-generation wide-area surveillance platform is intended to provide advanced integrated ground and air surveillance targeting and capabilities, and will detect, classify, characterise and report cruise missiles and surface targets to the joint forces. It will be based on the Boeing 767-300 airframe.

Venezuela

US down-plays MiG-29 plans
Press reports late last year that Venezuelan President Hugo Chávez was planning to acquire 50 Russian MiG-29s, risking a new arms race in Latin America, were played down in Washington by the senior State Department official for Latin American

affairs, Roger Noriega. He said that there were other countries in the region with those kinds of weapons, apparently referring to previous Soviet arms purchases by Cuba and Peru, which led him to believe that Venezuela's plans would not necessarily represent a military escalation.

FAV requirements were reportedly quoted as 40 single-seat MiG-29SMTs and 10 two-seat MiG-29UBTs in a comprehensive package which would cost up to $5bn, and include Vympel R-73 (AA-11 'Archer') close-combat agile IR-guided AAMs and R-27 (AA-10 'Alamo') medium-range radar-homing AAMs, plus air-to-surface weapons. Venezuelan defence budget restrictions would prevent outright purchase of the proposed Russian arms, for which an initial payment of up to $500m or so would be followed by trade agreements for supply of commodities and raw materials, such as aluminium, for the balance.

The wisdom of these proposals has also been queried in Washington, where it has been pointed out that the FAV funding problems mean that it can barely keep its 22 Lockheed Martin F-16s airworthy, and that its most pressing requirements are to replace its dozen or so remaining Rockwell OV-10A Bronco light ground-attack aircraft, supported by EMBRAER EMB-312 AT-27 Tucanos and the first of a dozen Brazilian-built two-seat AMX-Ts now being delivered, for close-support and drug-interdiction roles.

Helicopters delivered
The Kazan Helicopter Plant delivered three Mi-17V5 helicopters to the Venezuelan Army at its facility in Russia on 21 December 2005. Venezuela agreed to purchase six Mi-17V-5, three Mi-35M and one Mi-26T from Russia's Rosobornexport State Corporation during March 2005.

AIR ARM REVIEW

Australia

Tiger enters AAA service
Deliveries starting in December of the first two of 22 Eurocopter Tiger armed reconnaissance helicopters on order for Australian Army Aviation, at a ceremony at Oakey, in Queensland, marked achievement of the Project Air 87 In-Service Date (ISD) demonstration. Tigers ARH1 and 2 were manufactured in France, and delivered to Eurocopter's Australian Aerospace (AA) subsidiary in Brisbane on 23 November 2004. Australian Aerospace finalised the re-assembly and flight testing of the two platforms before transfer to the Army for formal acceptance. ISD achievement will lead to Australian Military Type Certification later this year.

Only the first four Australian Tigers are being completely built in the first quarter of 2005 by Eurocopter in France, where Thales is also producing the ARH aircrew simulators. The remaining 18 ARHs are being assembled at the Australian Aerospace production facility at Brisbane, where ARH 5 was the first to be ready for flight in late 2004, for delivery ahead of schedule this spring. The Army would be able to conduct tactical reconnaissance and escort, and protect its Sikorsky Black Hawk helicopters as they transport troops and supplies.

Canada

Flight training changes
The Department of National Defence (DND) has awarded a $1.5 billion flying training contract to Kelowna Flightcraft. Under the terms of the long-term, 22-year contract, Kelowna will provide flying training and support services to the Canadian Forces. Operations will be conducted at the Canada Wings Aviation Training Centre at the Southport Aerospace Centre, Portage la Prairie, Manitoba. Kelowna leads a consortium known as Allied Wings, which will provide primary and specialized helicopter and multi-engine fixed wing flight training.

Germany

US Phantom training ends
Retirement by the USAF of its last operational F-4 Phantom II unit on 20 December last year, with formal inactivation at Holloman AFB, New Mexico, of the 20th Fighter Squadron, known as the 'Silver Lobos', ended a 33-year German-American joint fighter training programme on the F-4E/Fs. They are still operated in first-line service by the Luftwaffe, after extensive systems upgrades, although replacement of its F-4Fs by Eurofighters is now beginning, ending German Phantom training requirement in the US.

When their assigned carrier, USS Theodore Roosevelt, put into port in Dubai for a few days, five F/A-18Cs from CVW-8 were deployed ashore to Al Asad AB in Iraq to continue combat operations. This VFA-15 aircraft taxis at Al Asad carrying AIM-9X, AGM-65, GBU-12 and JDAM.

India

Naval aviators train in Pensacola

Pilots destined to operate Russian-built MiG-29K fighters from the Indian Navy's new aircraft-carrier will be trained by the US Navy at NAS Pensacola, Florida. Four pilots who are already undergoing training began carrier qualification flights in January. A second group of four will commence the six-month course in March and a total of 32 aviators will be trained by the time the former Russian carrier *Gorshkov* arrives in India in 2008. After receiving their carrier qualifications the aviators will undergo MiG-29K transition training in Russia. The Indian Navy will operate 16 MiG-29Ks from the carrier.

Iraq

Air force re-formation continues

Reconstitution of the Iraqi air force, initially mainly for transport and support roles, received its biggest boost in December, from large-scale orders to Poland, via its Bumar arms agency, for some 44 helicopters. These include 20 twin-turboshaft PZL Swidnik W-3W Sokols, developed jointly with Russia's Mil OKB from 1970, and worth about $120m. Twelve are being configured with weapons systems for ground-attack roles, four equipped for search-and-rescue (SAR), and four with VIP interiors for official transport tasks. Iraq's Polish arms orders were also reported to have included two dozen Mil Mi-17 transport helicopters, presumably from military storage stocks.

IrAF helicopter strength was further boosted in February from delivery to Taji Air Base of the first two refurbished Bell UH-1H utility helicopters of 16 being transferred from Jordan through aid programmes. Some 14 Iraqi pilots have so far been trained and are awaiting additional UH-1 flight instruction from their US advisory support team (AST). Flight training will continue for the next several months until all 48 Iraqi pilots are certified, to staff Nos 2 and 4 IrAF Squadrons, each with eight UH-1s, from Taji. Maintenance training has also started for engineers and ground crews.

Deliveries have also continued at the rate of two per month until March to 70 Squadron of the reconstituted Iraqi air force in Basrah, of eight two-seat Zenith CH2000 Sama light surveillance aircraft. These were ordered through US aid programmes from AMD Aircraft in the

US, via Jordan Aerospace Industries (JAI) in Amman, from a $5.8m contract. Equipped with forward-looking imaging infra-red (FLIR) sensors and wide-bandwidth advanced data-link communications systems, the 86.42-kW (116-hp) Textron Lycoming O-325N2C-powered two-seat CH2000s were said by the IrAF Commanding General, Maj Gen Kamal Al-Barzanjy, to represent the second stage of rebuilding Iraqi military air strength.

They will be used for day/night surveillance and support operations, and an option to acquire another eight aircraft, on a similar monthly schedule, is under consideration. Meanwhile, operations will continue from Basrah, co-ordinated with coalition and Iraqi forces, with planned support from Kirkuk when a second squadron is formed early next year.

Ivory Coast

Air force to be resuscitated?

Having reportedly destroyed the entire Ivory Coast air force on the ground at Yamoussoukro Airport, in Abidjan on 6 November after two of its Sukhoi Su-25UBKs had attacked and killed nine French military personnel and injured 23 other people near Bouaké, the French and UN peacekeeping missions authorised the repair in January of four Su-25s and several of the helicopters which were not too badly damaged. Considering that French marines fired an Aérospatiale MATRA Roland short-range surface-to-air missile with a 6.5-kg (14.3-lb) pre-fragmented warhead at each parked aircraft to disable it, it seems surprising that any are capable of restoration.

Nevertheless, following renewed attacks in the rebel-held north, foreign technicians have been authorised to begin salvaging what they can from the Su-25s, two ex-Botswana BAe Strikemasters, four Mil Mi-24Vs, several Mil Mi-17s, and four IAR.330 Pumas. Spares acquisition may also be difficult, since the recent UN Security Council imposition of an arms embargo on the Ivory Coast.

Mexico

Fighters considered

The Mexican navy is considering expanding its aviation element and could purchase as many as eight fighter aircraft that would be configured for air defence and maritime strike duties. Mexico is considering the fighter purchase in response to concerns

Wearing temporary UK serial ZJ992, Merlin M503 for the Danish air force is seen undergoing trials at AgustaWestland's Yeovil factory.

over its ability to protect key areas such as oil fields located in the Bay of Campeche. Although the service previously looked at the procurement of the Aero Vodochody L-159, it is now reportedly looking at more capable aircraft, including the Saab Gripen and the Sukhoi Su-27. The navy is currently modernising both its fixed- and rotary-wing fleets and has recently taken delivery of the fourth of eight EADS/CASA C-212s, which are configured for counter-drug and SAR duties. Additionally, its fleet of 11 MBB Bo105 helicopters is being upgraded and the Navy plans to acquire five new Mi-17TV1 transport helicopters.

Poland

First Polish F-16 pilot qualified

The first Polish Air Force pilot destined to fly the air force's new F-16C/D fighters recently completed a seven-month transition with the Arizona Air National Guard's 162nd Fighter Wing at Tucson IAP. In total the unit will train 48 pilots over the next three years and 13 have already begun or completed the course, which includes 62 flights. Poland will take delivery of the first of 36 Block 52 F-16Cs and 12 F-16Ds during 2006.

Qatar

Mirage 2000s withdrawn?

According to press reports, agreement was reached late last year between the Doha and Indian governments for acquisition by the IAF of nine Qatar Emiri air force Dassault Mirage 2000-5EDAs and two two-seat Mirage 2000-5DDAs. With another Mirage 2000-5DDA, these were delivered to the QEAF only in 1997, but their operation was not economically justifiable in the absence of a realistic threat. Some reports claim that Indian interest in the QEAF Mirages was stimulated by

earlier negotiations between Qatar and Pakistan for their possible acquisition. The QEAF will now have no combat element, apart from any weapons that may be carried by its half-dozen Alpha Jet advanced trainers.

Singapore

Apache news

The Republic of Singapore Air Force shipped the first three of its AH-64D helicopters to Singapore early in February. The RSAF's five remaining Longbow Apache helicopters from its initial order of eight stay based at Silverbell Army Heliport in Marana, Arizona, operating alongside the Arizona Army National Guard's 1-285th Aviation Regiment. Singapore's Apaches recently made their debut at an integrated exercise known as Forging Sabre, which was conducted at the US Marine Corps Air Ground Training Center at Twentynine Palms, California, and involved RSAF CH-47D helicopters and F-16 fighters that are all stationed in the US. The Marana-based examples are operated by the RSAF's Peace Vanguard Detachment. The RSAF accepted the first of 12 additional helicopters in January and its fleet of 20 Apaches is expected to be fully operational in two years.

Thailand

Service entry for Super Lynx

Formal acceptance aboard HTMS *Taksin* took place by Royal Thai Navy C-in-C Admiral Sampop Amraparn at Sattahip Naval Base on 7 February of the first AgustaWestland Super Lynx 300 to enter RTN operational service. Ordered in late 2001, the new Super Lynx will be operated by No. 203 Squadron from U-Tapao when land-based, for a wide range of naval roles including maritime patrol, search and rescue (SAR), and ASW/ASuW. The

The Swedish Air Force retired the Viggen in December 2005, and several aircraft were rapidly dispatchged to museums. This former 1. Div/F21 AJSH 37 is seen on arrival on 13 January at Lelystad in the Netherlands for display at the Aviodrome museum.

YI-301, one of three C-130Es assigned to the Iraqi Air Force's 23rd Squadron, arrives at its new base of New Al Muthana AB in January 2006 having relocated from Ali AB. The squadron flew its first mission with an all-Iraqi crew on 28 November 2005.

On 25 November 2005 the Austrian Air Force officially celebrated the retirement of its Saab J 350e Drakens. For the last Draken solo display '21' received a special 'Dragon Knight' paint scheme in black and yellow, with the flasgs of the air arms that flew the type on the fin. Seven Drakens remained flying to the end of 2005. In 18 years of service the Austrian double deltas logged more than 23,545 hours in the air without loss.

aircraft will mainly fly from the RTN's 'Narasuan'-class frigates, but can also operate from all other Royal Thai Navy ships with aviation facilities.

United Kingdom

Cobham contract renewed

FR Aviation (FRA), part of Cobham's Flight Operations and Services Group, has secured a £140m ($263.2m) five-year extension from 2009 to 2014 of its 20-year partnering arrangement with MoD to provide aerial support and training services. They involve some 7,000 annual flying hours by 15 specially-equipped FRA-operated twin-turbofan Dassault Falcon 20 light transports.

They provide essential highly-specialised electronic warfare training, threat simulation, silent target and aerial target-towing in the UK and overseas for the RAF and RN, in support of national, multi-national and NATO exercises. With more than 500,000 flying hours of civil and military aircraft expertise, FRA aircrews can deliver sophisticated effects-based warfare training, directly contributing to the safety and effectiveness of UK defence forces personnel. Cobham's Flight Operations and Services Group operates over 150 aircraft world-wide on military training, special mission flight operations, freight and passenger services and large military aircraft maintenance.

United States

Final Sea King overhaul

The final UH-3H helicopter overhauled by IMP Aerospace was delivered to NAS Patuxent River, Maryland, from the

On 1 March 2006 the RAF's No. 45 (Reserve) Squadron celebrated its 90th anniversary and also two years of flying the Beech King Air 200 in the multi-engined pilot training role. To mark the event ZK453 'M' received a commemorative fin. At the beginning of 2006 five of the squadron's seven aircraft were taken onto the military register to comply with the provisions of the 1947 Chicago Convention involving civilian-owned aircraft being operated by military personnel over member states.

contractor's facility in Halifax, Nova Scotia, Canada recently. The contractor carried out standard depot-level maintenance (SDLM) overhauls on 25 US Navy Sea Kings since 1999. The air station's SAR unit currently operates four UH-3Hs that were overhauled and refurbished by IMP Aerospace. The programme provides the helicopter with a service life extension of 2,000 flight hours, which is equivalent to about six years of service. The navy currently plans to retain the Sea King in service through 2010.

'Saints' fly south

Composite fighter squadron VFC-13's Detachment Key West recently conducted its first operational tasking in support of Strike Fighter Squadron VFA-106 Super Hornet training at the Florida Naval Air Station. The 'Saints', which will operate 12 F-5N Tiger II fighters, flew 158 adversary training sorties over a two-week period with just eight aircraft. The detachment's 20 aviators include eight on active duty and 12 Naval Reservists. Although activated in October 2005, hurricanes resulted in several evacuations and damage to the south Florida base, delaying the unit's operational debut until 2 December 2005, when the first sorties were conducted. The main section of VFC-13 is based at NAS Fallon, Nevada.

MH-60R fielded

The Navy's newest helicopter arrived at NAS North Island, California, on 5 December 2005, when helicopter antisubmarine squadron light squadron HSL-41 took delivery of its initial pair of MH-60R Seahawks. The multi-mission MH-60R will operate both from aircraft-

carrier flight decks and those of smaller vessels that currently support the SH-60B. As part of the transition process, the squadron was subsequently redesignated as Helicopter Maritime Strike Squadron HSM-41.

Viking fun4eral
The 'Diamondcutters' of sea control squadron VS-30 were deactivated at NAS Jacksonville, Florida, during ceremonies held on 9 December 2005. The squadron, which operated the S-3B as a component of Carrier Air Wing (CVW)-17, completed its last deployment aboard the USS *John F. Kennedy* (CV 67) during February 2005. It flew its last operational mission during September when it conducted a joint counter-drug operation with VS-31 and the USAF's 429th Expeditionary Operations Squadron in the Netherlands Antilles.

Navy Reserve drawdown

As part of the planned drawdown of its P-3C fleet, Navy Reserve patrol squadrons VP-66 at NAS JRB Willow Grove, Pennsylvania, VP-65 at NAS Point Mugu, California, and VP-94 at NAS JRB New Orleans, Louisiana, were deactivated on 31 March 2006. The latter squadron had been operating at NAS JRB Fort Worth, Texas, since relocating from its home station, in the face of Hurricane Katrina in September 2005. The reserve's three remaining squadrons will be retained but each will operate just six Orions.

'River Rattlers' return

Strike fighter squadron VFA-204 returned to NAS JRB New Orleans on 13 December 2005. The squadron had been operating from NAS JRB Fort Worth, Texas, since evacuating the Louisiana base in the face of Hurricane Katrina in September 2005.

Carrier updates

Updating its announcement to replace the USS *Kitty Hawk* (CV 63) with a nuclear-powered aircraft-carrier, the Navy has confirmed that it will transfer the USS *George Washington* (CVN 73), currently based in Norfolk, Virginia, to Japan in 2008.

Sea Knights retired

Marine medium helicopter squadron HMM-162 stood down in preparation for its transition to the MV-22B on 9 December 2005. Based at MCAS New River, North Carolina, the 'Golden Eagles' will be the second operational Osprey squadron when it stands up again as Marine medium tiltrotor squadron

VMM-162 in November 2006. The squadron's 12 CH-46Es were transferred to other units. Approximately half of HMM-162's personnel were transferred to VMMT-204 while the remainder were assigned to other Sea Knight units.

Space Command reorganised

Four Helicopter Flights (HF) assigned to Air Force Space Command (AFSPC) were recently redesignated as squadrons. The units, which operate UH-1N helicopters, are primarily tasked with supporting intercontinental ballistic missile (ICBM) squadrons. This includes missile field security, wartime contingencies, training and SAR. Additionally, they are tasked with conducting aerial surveillance when nuclear weapons are moved, emergency deployments of security response forces throughout the base and missile complexes, and priority airlift of both cargo and passengers. Prior to being redesignated the four flights deployed in support of relief efforts associated with Hurricane Katrina. In total, Air Force Space Command (AFSPC) deployed eight UH-1Ns and crews to Columbus AFB, Misssissippi, during September 2005. The deployment marked the first time the AFSPC helicopters were deployed from the missile fields and assigned to tasks outside their normal missions. The assets, which were assigned to the 620th Air Expeditionary Squadron (AES), comprised two helicopters from each of the flights and 83 flight and maintenance personnel.

Wing	Sqn	Location
30th SPW	76th HS	Vandenberg AFB, Calif.
90th SPW	37th HS	F.E. Warren, Wyoming
91st SPW	54th HS	Minot AFB, North Dakota
341st SPW	40th HS	Malmstrom AFB, Montana

Globemaster III moves

The Hawaii Air National Guard's 154th Wing flew its last C-130H sortie on 24 January. The wing's 204th Airlift Squadron had operated five examples of the airlifter since 1990, and subsequently turned in serial 90-1058, its final aircraft and transitioned to 'associate' status. It will now share eight C-17As assigned to the 535th Airlift Squadron, which is an active duty unit assigned to the 15th Airlift Wing. Both wings are stationed at Hickam AFB, Hawaii. The commander of the Pacific Air Forces (PACAF) formally accepted the first C-17A assigned to the 15th AW/154th Wing's in ceremonies held at Boeing's Long Beach, California, facility on 7 February. Nicknamed *Spirit of Hawaii Ke Aloha,* the Globemaster III was subsequently flown to MCAS Kaneohe Bay, Hawaii, and delivered to Hickam AFB the following day where it was formally christened before entered service with the 535th and 204th AS.

Air Mobility Command's 60th Air Mobility Wing (AMW) and the Air Force Reserve Command's 349th AMW are making the final preparations for the arrival of the first C-17A at Travis AFB,

Carrying markings for the 249 missions it undertook while on deployment to Al Dhafra in the UAE, RQ-4 AV-3 (98-2003) lands back at Edwards AFB, California, on 20 February 2006 after three years supporting Central Command operations in Afghanistan, Iraq and the Horn of Africa. The aircraft is assigned to the

Reserve Command's 932nd Airlift Wing at Scott AFB, Illinois. In preparation for the 'new' aircraft, the 932nd similarly transferred its last C-9A to storage at the Aerospace Maintenance and Regeneration Center (AMARC) at Davis Monthan AFB, Arizona.

Deployed squadron relieved

The 28th Bomb Wing's 34th Bomb Squadron (BS) relieved the wing's 37th BS that had been deployed to Andersen AFB, Guam, in late December. The latter squadron had deployed to the Pacific base from Ellsworth AFB, SD, along with six B-1Bs earlier in 2005. The deployed aircraft and crews are attached to the 36th Air Expeditionary Wing at Andersen.

Air Force Academy changes

The 70th Flying Training Squadron (FTS) has been activated at the USAF Academy in Colorado Springs, Colorado. The unit,

which is a component of the Air Force Reserve Command (AFRC), had previously been known as Detachment 1, 302nd Operations Group. The unit has provided instructors to the 306th Flying Training Group (FTG) since June 2004, augmenting the group's 94th, 98th and 557th FTS. The squadrons respectively provide cadets with instruction in flying gliders, freefall parachuting, and operate the flight screening programme and support the cadet flying team. The 70th FTS was last active with the T-37B as part of the 38th Flying Training Wing (FTW) at Moody AFB, Georgia. It was inactivated in December 1975.

UAV test squadron

The 30th Reconnaissance Squadron has been activated at the Tonopah Test Range, Nevada, and assigned to the 57th Operations Group at Nellis AFB. It is unclear as to what aircraft types are operated by the squadron, which will reportedly be tasked with operational test and evaluation (OT&E) duties. Last serving as a component of the 10th Tactical Reconnaissance Wing, the 30th Tactical Reconnaissance Squadron flew RF-4Cs before being inactivated at RAF Alconbury, UK in April 1976.

Apache training ends

The US Army Aviation Center at Fort Rucker, Alabama, said farewell to the last example of the AH-64A attack helicopter recently. Training for the 'A' model had been carried out by A Company, 1-14th Aviation Regiment at Hanchey Army Airfield. The unit's final graduates included personnel assigned to the Army Reserve Command's 8-229th Aviation at Fort Knox, Kentucky, and the 7-6th Aviation in Conroe, Texas. All training associated with the AH-64A is now conducted by the Army National Guard's Western Army Aviation Training Site (WAATS) in Marana, Arizona.

Army National Guard changes

The Florida Army National Guard's 1-111th AVN recently took delivery of the first of six CH-47Ds at the Cecil Commerce Center in Jacksonville. The general support battalion's three companies, which previously operated the AH-64A, are transitioning to new missions and aircraft, including the Chinook and UH-60 Blackhawk. The latter will serve both in general support and air ambulance roles.

Schofield changes

The aviation brigade of the 25th Infantry Division (Light) at Schofield Barracks, Hawaii, has commenced a reorganisation, in support of US Army Aviation transformation efforts, that will result in its being reflagged as the 25th Combat Aviation

Brigade. A new general support aviation battalion was activated as 3-25th Aviation and it will eventually have three companies equipped with Blackhawk and Chinook helicopters. Additionally, the 3-4th Cavalry, which operated the OH-58D, has been reflagged as the 6-17th Cavalry. The final major change, which is scheduled for mid 2006, will result in the reflagging of the 1-25th Attack Helicopter Battalion as the 2-6th Cavalry.

Blackhawks to Sinai

On 2 December 2005 five UH-60As were loaded into the cargo compartment of an Antonov An-124 operated by Volga-Dnepr Airlines at Pope AFB, North Carolina. The aircraft were subsequently flown to Egypt where they were assigned to the 1st US Army Aviation Support Battalion Aviation Company. The unit is a component of the Army's 1st Corps Support Command, which is also known as Task Force Sinai. The Blackhawks replaced UH-1H/Vs operated by the unit, which supports the Multinational Force and Observers in the Sinai. Three additional Blackhawks were shipped to the company, which is based at Al-Arish Airport, El-Gorah, aboard a USAF C-17A on 30 January.

Homeland Security changes

On 1 October 2005 US Customs & Border Protection (CBP) integrated the aviation assets assigned to the Office of Border Patrol and the Office of Air & Marine Operations to form a new organisation known as the Office of CBP Air. The organisation was renamed again, becoming CBP Air & Marine on 18 January 2006,

when CBP's marine assets were rolled in. CBP Air & Marine, which is the largest law enforcement air force in the world, is responsible for more than 250 fixed- and rotary-wing aircraft, and has over 500 pilots. The organisation is headquartered in Washington, D.C.

Washington air defence

The responsibility for intercepting potential threats in the air space around Washington D.C. and the National Capital Region (NCR) will be transferred from the US Customs and Border Protection's CBP Air & Marine to the US Coast Guard later this year. The 5th US Coast Guard District, headquartered in Portsmouth, Virginia, will be responsible for the air defence mission. The actual duties will be assigned to five HH-65C helicopters operated by Air Station Atlantic City, New Jersey. Three of the helicopters will be forward-deployed at Ronald Reagan National Airport, in Arlington, Virginia. The airport is currently home to the UH-60A Blackhawk helicopters used by CBP Air & Marine to carry out the mission.

In order to fully support this duty, the Coast Guard will acquire seven additional HH-65Cs, comprising two that will be purchased in FY07 and two that will follow in FY08. The service's proposed budget for FY07 includes $62.4 million to establish the National Capital Region air defence operation in Washington, including $48 million that will cover the purchase and upgrading of the first batch of new HH-65s. The Dolphins will eventually be armed as part of the coast guard's Airborne Use of Force (AUF) programme.

The Israel Air and Space Force (IASF) is repainting its Bell 206 Saifan (Avocet) fleet of primary rotary-wing trainers in a variation of the IASF Flying School's white scheme with red trim. Elbit Systems subsidiary Cyclone is repainting the helicopters and the process is to be completed by the end of 2007. Cyclone has been awarded an Israeli Ministry of Defence (IMoD) contract to maintain the IASF Flying School's training helicopters under a 10-year 'power by the hour' (PBTH) agreement, announced on 2 June 2004. Saifan 126 is the first to be repainted in the white and red scheme.

F-22A

In service with the 1st FW

After a long development, the USAF's latest fighter has entered service with the Langley-based 1st Fighter Wing's 27th Fighter Squadron. While debates still surround the remainder of the Raptor programme, the unit is setting about the business of learning how to use a fifth-generation fighter, and putting that knowledge into practice. Already the 27th FS is regularly demonstrating just how superior the F-22A is when compared with existing fighters.

The cockpit of the F-22A is the most sought-after in the Air Force. The selection process is rigorous and carefully planned, so that the growing F-22 community has the right balance of young and more experienced pilots. Above-average skills are a pre-requisite, no matter what the level of experience.

It is only fitting that the First Fighter Wing based at Langley Air Force Base, Virginia, has become the first active duty wing within the United States Air Force to stand up with the F-22A Raptor fighter. Even more appropriate is the fact that the 27th Fighter Squadron, the 'Fighting Eagles', is the first active fighter squadron to operate the Raptor. The 27th FS is the oldest fighter squadron in the USAF, dating back to 1917. Initially designated the 27th Aero Squadron, the unit began flying the SPAD and Sopwith Camel aircraft. Things have changed a bit since the unit flew cloth-covered aircraft. The 27th FS was the first active duty unit in the Air Force to take receipt of the Eagle back in 1976, and is now trading in its current jet for the Raptor.

If there is such a thing as a fifth-generation fighter, the F-22A Raptor is it. Designed to replace the excellent F-15 Eagle in the air superiority role, the Raptor is a quantum leap forward in fighter aircraft technology and capability. Blending a combination of stealth, advanced avionics and engines capable of supersonic flight without the use of afterburners, the Raptor is definitely next generation. In mock combat missions a consistently outnumbered F-22A devastated F-15s. This is all the more impressive when one considers that, up until this point, the Eagle has amassed a perfect record in combat, destroying over one hundred opponents without a single loss.

IAPR had the opportunity to speak with the Director of Operations for the 27th FS, Lieutenant Colonel Wade Tolliver, a 16-year veteran of the USAF hailing from Kissimmee, Florida. The Colonel entered the Air Force as a Reserve Officer Training Candidate while attending the University of Florida. Going by the call-sign 'Troll', Lt Col Tolliver has amassed 2,200 hours in the F-16 and just under 100 hours in the F-15C.

A pair of 27th FS F-22As formate during a training sortie. The squadron undertook its first operational sortie in January 2006, joining other fighter units on the Homeland Defense mission.

F-15C Eagles are still much in evidence at Langley, although the second squadron (94th FS) began re-equipping with the F-22A in March 2006. The last of the three 1st FW units to convert will be the 71st FS, which will fly the F-15 for a few more months.

The colonel relates an interesting story from a recent flight in which he was involved in pitting Raptors against Eagles, illuminating the vast differences between these two warbirds. "We took out a two-ship of F-22s and we were employed against four F-15Cs, also from the 1st FW. My wingman was 30 miles [48 km] in trail as we approached the division of F-15s. The basic plan was for him to come in behind me and clean up any Eagles that I missed. I engaged the four Eagles while travelling at 50,000 ft [15240 m] and Mach 1.6, and I ended up killing all four, quickly leaving my exasperated wingman asking why I didn't leave anyone for him to attack. Anytime you can go against the F-15C – which is the finest air superiority fighter in the world – and you can have results like these you must be employing something amazing.

"There have been radio calls in previous flights where the Eagle pilots ask over the radio if we are even out there. They never see us and they can't engage what they can't see. It's pretty frustrating for them; these are great pilots in great airplanes getting shot before they can even engage. It's not a lot of fun and, as you can imagine, sometimes I have trouble getting guys to volunteer to serve as our adversaries. It's unfair, but that's the way we want to keep it, and that's the way it should be for the United States to insure total air dominance. We need to maintain this unfair advantage so our naval forces, our ground and Special Forces can operate safely and effectively."

Overseeing operations

In his current assignment, Lt Col Tolliver is responsible for all flying operations for the unit. "The challenges in this assignment are many, especially considering the 27th is the first unit with this amazing new aircraft. The Director of Operations oversees all the training requirements for all the pilots. This includes ground and flying training, and everything necessary to get the guys ready for combat. As you can imagine, in standing up a new fighter squadron with the F-22 my concern is to get the guys ready to go to combat in an aircraft that has never been in combat. The challenge with that is that the F-22 cannot be thought of as a legacy fighter.

"The capabilities of this aircraft are such that we need to think outside the box on how we employ it and the way train. The Raptor is so transformational, so different from anything we have ever flown, we can't think like we did in the past about how we get a pilot ready to go. So, the big challenge I have is to understand how are we going to employ this aircraft when we get into combat? And number two, how do I get the guys ready to do that. The second part of the equation is, with these new aircraft coming into the squadron, we have to make sure they are able to fly. Our maintenance folks have done an outstanding job keeping these aircraft fully mission-ready. As with any new system we are meeting new challenges along the way but we are dealing with them and getting better at it."

Mission Ready syllabus

There is a formal training syllabus that is used to train the pilots and get them the proficiency they need to be ready to go to combat. Lt Col Tolliver comments, "We have an in-depth training programme to get a pilot proficient to serve in combat and be considered Mission Ready. For him to be ready he finishes the initial training at Tyndall AFB and now has the basic proficiency and has seen every type of mission the Raptor faces by the time he returns to us. We take that basic proficiency and turn it into a much higher level of proficiency so he is ready to take this

On the ramp at Langley, a 27th FS F-22A is seen with its weapon bay doors open. The side bays carry Sidewinder missiles, mounted on a rail that deploys outwards (at an angle) so that the missile can fire straight off the rail without damaging the aircraft.

airplane anywhere in the world, and employ it within any mission he is called upon to do. We have the ability to modify the syllabus based on how the pilot did during the initial training: if he had trouble in certain areas we provide additional time and training so there is not a rigid 'cookie-cutter' approach. We make sure we give each individual everything needed so they are ready to go to war."

The cockpit of the F-22A is highly sought after with many qualified candidates from the ranks of the Air Force's fighter community. The unit did not simply swap aircraft but was actually rebuilt anew from the ground up. On 1 October 2004 the unit officially became an F-22A Raptor unit (at the time the aircraft was designated F/A-22 to reflect its attack capabilities. This was changed to the current F-22A in December 2005). The 27th is one of three flying fighter squadrons comprising the First Fighter Wing (1st FW). The other two units are the 71st and the 94th Fighter Squadrons. Both the 71st and 94th are upgrading to the Raptor, the 94th receiving its first two aircraft in March 2006.

According to Tolliver, "We will ultimately have 26 F-22As, and we'll have a total of 32 pilots. Four will be assigned to the wing; the others will be in the unit. There are approximately 180 people assigned to the 27th Aircraft Maintenance Unit, supplemented by numerous other people throughout the 1st FW that have the responsibility of maintaining this aircraft."

Lieutenant Colonel Tolliver went on to discuss the transition process, "When we made the transition we transferred all of our pilots and all of our aircraft to the other two units here at Langley. There were three pilots that were assigned to the 27th initially, the squadron commander, one other pilot and me. We had no airplanes but we did have enlisted troops that were in the Operations side of the house, as well as the maintainers that were training to work on the aircraft.

Initial cadre

"As time progressed and we prepared for the arrival of our first aircraft, which was slated for January, we began selecting the initial cadre of pilots that would make up the squadron. Those pilots were hand-picked and we put a lot of effort into which ones were being selected to join the squadron. As you can imagine, with the importance of a programme of this nature, we wanted to give it the best opportunity to succeed. We went out and picked the best possible guys, but we did so within specific demographics.

"We did not want to just pick a bunch of older guys like myself. We needed some younger captains, guys that could grow up in the F-22A, so we selected a mix of younger and older guys. There were a lot of constraints to make sure we had the right demographic. There are many pilots out there that are great fighter pilots that deserve to fly this aircraft. We couldn't take every capable pilot in the Air Force, so we selected those that we thought could do the best job that also met the demographics and, of course, their career timing had to work out for them as well."

For those familiar with the USAF fighter community, the elite aircrews can often be found in the ranks of those that have graduated from the prestigious Weapons School, located at Nellis AFB, Nevada. This demanding graduate-level programme is over six months long and is designed to take a young aircrew selected by their

As well as its primary air defence role, the F-22A will be used for attack duties, including close air support. The primary weapon will be the GPS-guided JDAM.

These F-22As display the unusual specular paint applied to the aircraft. Noteworthy on the aircraft that is peeling away are the large, blended fairings covering the mainwheels and the actuators for the ailerons, flaps and tailerons.

commanding officers and make them experts in the employment of their aircraft. Upon graduation, these officers, referred to as 'Patchwearers', are some of the most highly trained flyers in the USAF.

This is verified by Tolliver: "The perfect prototype that we were looking for, even dating back to the beginning of the Raptor test programme at Edwards and Nellis, were 'Patchwearers'. We were fortunate enough to bring in some 'Patchwearers' to the 27th that did some of the initial testing, and they brought a wealth of experience with them.

"For the younger pilots we were interested in guys that worked very hard in their squadrons at being the best they could be, whether they were in the F-16, F-15C or F-15E. We were looking for a person that knows their job, the tactics and is very good at employing fighter aircraft. Those that

were selected were of the calibre that, had they stayed in their unit, they would have gone on and been selected to go to the Weapons School anyway."

While it can be said that the 27th FS is selecting the cream of the crop, Lieutenant Colonel Tolliver is careful to note that, "we are being very careful about this selection process and keep the interests of the entire USAF in mind. We want to ensure we do not pillage the rest of the other fighter communities and take the best of their people. With the number of Raptors the US is buying, there will still be many F-15C, F-15E and F-16 units that will need to be manned with solid fighter pilots. We are not going to go out and take all the best young guys from these other weapon systems: this is a big Air Force and we need those smart guys out in their respective communities."

Flying the Raptor

Tolliver describes his impressions of flying the world's newest and most lethal warplane: "They did not build any two-seat Raptors. The first time

you sit in the airplane and take off you are by yourself. More importantly, the first time you land it you are by yourself. The simulators today are of such high fidelity that, when I first got in the F-22, I felt very, very comfortable. I wasn't worried about taking off or landing. I got to the end of the runway, and you're always a little nervous, it's your first flight in a new airplane, you don't want to mess anything up. The vision that went through my head that made me laugh a bit was that I was not worried about my personal safety; I was worried about scratching this new airplane that cost so much. When I look back on my first flight the aircraft took off and landed just like the instructors taught me it would, and just as it did in the simulators. I felt very comfortable doing it and I had no problems during my first flight whatsoever.

The secret to stealth is maintaining the 'LO bubble' – that is, the clean shape of the aircraft. For missions where low observability is not required – notably for long-range ferry flights – the F-22A can carry stores and tanks on wing pylons.

"The pre-mission planning is a little bit different when comparing the F-15 to the F-22A. Going out to the airplane itself and pre-flighting the aircraft is a bit different. All the weapons are carried internally, unlike the F-15 that has them all externally. That was a bit different, having to bend down and look into the weapons bays to see the weapons. As far as taxiing goes the Raptor is a little noisier inside the cockpit than the F-15, it creaks a bit like a big ship. The noises it makes got my attention, as well as the sensitivity of the brakes. They are a great design and they work well. I was a little bit jerky the first time I taxied since I was not used to the sensitivity of the brakes."

The Raptor is roughly comparable in size to the F-15 although it gives one the impression initially of being larger. It's a bit beefier in the middle and carries a similar paint scheme to those worn by Eagles. While elegant in the air, the Raptor is a bit less so on the ground. Its short landing gear seems almost too small for the airplane. Lt Col Tolliver provides some insight on this: "This aircraft was built with the maintainer in mind. The landing gear was designed intentionally to help the maintainers. The F-15C sits so high atop the landing gear, the maintainers have to use a ladder to do most of the work on the aircraft. The F-22A sits lower, making it easier to work on. The Raptor is a foot shorter in length than the F-15C and about a foot wider."

Multi-role aircraft

The Raptor was initially conceived two decades ago to be strictly an air superiority fighter, just as the F-15A Eagle was upon its introduction. In this day and age of booming budget deficits the Air Force saw the wisdom in affording the Raptor a multi-role capability. The aircraft is capable of using precision guided munitions like the JDAM, giving the F-22A a devastating air-to-ground capability and further enhancing its value to the American military. Lt Col Tolliver elaborates: "We are completing an aircraft modernisation that gives the Raptor the software capability to carry the JDAM. Now that we have this, we are focusing more and more on the air-to-ground mission.

Keep in mind we have always trained in the simulators in the air-to-ground role because we believe that's going to be our primary mission. In fact, next week we are doing a week-long CAS training exercise which includes simulating dropping JDAMs.

"This aircraft is totally different to the F-15 and F-16. It brings something very unique to the fight. There is simply no comparison. It's been described as unfair when we fight F-15s. The battlespace awareness you have in the F-22 is unlike any other airplane in the world. With this awareness, you can employ your weapons and meet your mission objectives so much easier because you have such a high level of knowledge as to what's going on around you. No other airplane in the world offers you that. The advantages of the F-22A are many – the stealth, avionics, supercruise.

"By themselves, these features stand out beyond what any other fighter can do. But with the Raptor it's a package deal, when you put all these things together you have created an incredible aircraft. The stealth alone, with the ability to not be seen to engage the enemy undetected is unprecedented. I can find him, shoot and kill him before he ever sees me on his radar. That alone, gives a significant advantage. With supercruise (which by definition means I can travel at supersonic speeds without using afterburners) gives me increased weapons ranges and keeps me exposed to threats for much less time than in an F-15 or F-16. I use less gas to maintain higher speeds and without being in afterburner I lower my heat signature to any infra-red threats. The integrated avionics are, in my opinion, the biggest plus in this aircraft. The situational awareness each pilot has in his cockpit – which by the way is able to be shared with other members of his flight – is superior to anything else in any other fighter.

"Focused logistics is another attribute of the Raptor. The Raptor requires fewer maintainers, and also requires less equipment which gives you the advantage of having less forward footprint in personnel and gear when deploying. It takes less aircraft to get you and your support elements to the fight. The Raptor is designed to be much more reliable as well, so down-time between flights is significantly reduced. When you take these attributes and roll them all together you have an incredible asset that can be employed anywhere in the world against 'double-digit' SAMs and defeat enemy aircraft from anywhere in the world, as well as those on future drawing boards."

Rick Llinares

The F-22A gives fighter pilots everything they could want in the modern era – manoeuvrability, stealth and a highly capable weapon system. That comes at a high price, however, and F-22 production has been incrementally cut as the programme has progressed.

The Gambia

Gambian Air Wing

Wedged into Senegal on the West African coast, the Republic of The Gambia is the smallest nation in Africa, with an area of 11295 km² (4361 sq miles) that follows the Gambia river inland. It gained independence from the United Kingdom on 18 February 1965, and later formed the short-lived federation of Senegambia with Senegal between 1 February 1982 and 30 September 1989. In 1991 the two nations signed a friendship and co-operation treaty. A military coup on 22 July 1994 overthrew the president and banned political activity, but a new constitution in 1996 and presidential elections, followed by a parliamentary election in 1997, completed a nominal return to civilian rule. The country underwent another round of presidential and legislative elctions in late 2001 and early 2002. Yahya Jammeh, the leader of the *coup*, has been elected president in all subsequent elections. The Gambian armed forces number around 1,900 and the 2004 expenditure was $1 million, or 0.3 per cent of GDP. There is a National Army, Navy, National Guard and Presidential Guard.

In June 2004 the Gambian Air Wing acquired its first and so far only aircraft in the form of a single Sukhoi Su-25 'Frogfoot'. This was acquired

Still wearing its New Millennium titles, Il-62M C5-GNM is operated by The Gambian government on VIP and presidential transport duties.

from the Georgian air force and was flown by Captain Peter Singahtey. It was delivered in its old colours and registration ('81' blue, c/n 10630) but subsequently the Georgian roundel was replaced by The Gambian flag. For most of the time it sits in a corner at Banjul-Yundum International Airport – the country's only airport – but has been flown on ceremonial occasions such as the 22 July anniversary of the coup.

Other government aircraft include an Ilyushin Il-62M. This aircraft (c/n 3036142, TL-ACL) was one of several operated by Centrafricain Airlines, controlled by notorious arms trafficker Victor Bout, an ex-KGB major. It was taken over in 1999 by Gambia-based New Millennium Air as C5-GNM, and occasionally used for presidential flights. New Millennium ceased operations in 2002 after the director, Babe Jobe, was listed under UN Resolution 1343 for arms trafficking to Liberia. The Il-62M was taken over by The Gambian government as a presidential aircraft, but retains its old titles.

The most recent government acquisitions are two newly-built Air Tractor AT-802A agricultural aircraft, each costing over $1 million. President Jammeh also acts as agriculture minister, and took delivery of the pair on 23 April 2005. The Air Tractors are used for a variety of purposes, including spraying operations against locust invasions, malaria and for fire-fighting. The pair are registered DOSA-1 (c/n 802A-0186, ex-N186LA) and DOSA-2 (c/n 802A-0195, ex-N554LA). With 3028-litre (800-US gal) hoppers the Air Tractors can cover a fair amount of The Gambia's territory in a day in good weather.

Dick Lohuis

Left: AT-802A DOSA-2 sits on the Banjul ramp next to the Su-25. The insecticide drum in the foreground points to the aircraft's primary role in Gambian service. The Air Tractors were ferried from the US via the Azores.

Below: This Su-25, The Gambia's sole military aircraft, has flown only sporadically since being delivered from Georgia in June 2004 but is maintained in good condition by the Air Wing at Banjul.

YF-117 retirement

A highlight of the 2005 Edwards AFB open house was the farewell flypast of YF-117 Nighthawk 79-10782, 'Scorpion 3', flown on 22 October by Lt Col Sean McAllum, commander of the 410th Flight Test Squadron (FLTS), and by Lt Col Dwayne 'Pro' Opella, 410th FLTS operations officer, the following day. Covering its entire underside was a red, white and blue American flag paint scheme applied to mark its retirement after 24 years of service. These patriotic colours were inspired by a similar scheme applied to 782 in December 1983, four years before the F-117's public unveiling, when it swept in low over a crowd that was attending the Joint Test Force change-of-command ceremony of Lt Col Roger Moseley to Lt Col Paul Tackabury.

Test history

YF-117 782 has been used for flight test ever since Lockheed pilot Tom Morgenfeld first took it into the air on 18 December 1981. As the third of five Full Scale Developmental (FSD) airframes, it was the original avionics testbed for the Nighthawk programme.

Five hand-built YF-117 prototypes (79-10780 to 10784), nicknamed 'Scorpion 1' to '5', were built prior to the type entering production. Ship 782, or 'Scorpion 3', was the first to have a complete avionics suite, enabling it to test full integration of the weapons delivery system. As such it was the first to be fully equipped with the Infrared Acquisition and Designation System (IRADS), which included Forward Looking Infrared (FLIR) and Downward Looking Infrared (DLIR) sensors. These two components allow the pilot to see at night and 'laze' the target for weapons delivery.

By April 1982, installation of the stores trapeze allowed it to become a weapons delivery testbed as well. Major milestones in stores separation tests occurred later in 1983, when 782 was used to drop the first full-scale weapon, a Mk 84, on 25 October, followed by a GBU-10 two days later. The first auto-separation took place on 17 December 1983.

By March 1993 the 410th FLTS was established to assume F-117 test duties. Between March and September, 782 (using the call sign 'Grey Ghost') received a new grey paint scheme under the Evening Shade programme, but just prior to the Edwards AFB Open House in October 1993 it was repainted black. It was never upgraded

The 'Stars and Stripes' filled the entire underside of 782 for its retirement flypast. Below 'Scorpion 3' is shown with another FSD YF-117, 784/'Scorpion 5' that remains in use with the 410th FLTS detachment at Palmdale.

beyond the second-generation technology improvements, thus becoming the 410th FLTS test platform for other F-117 fleet upgrades.

Retirement

After the F-117 Combined Test Force (CTF) had obtained approval to re-apply the flag design on 782 to commemorate its retirement, a group of CTF volunteers (Air Force and Lockheed Martin) led by Chief Master Sergeant Terrance McMahon

painted the flag scheme during October 2005. On 19 October, after emerging from the paint shop, 782 was taken up by Lt Col 'Grumpy' McAllum (410th FLTS Commander) for a photo session accompanied by YF-117A 784 piloted by Lt Col Opella.

Soon after the Edwards fly-over, the flag scheme was removed and the aircraft was prepared for retirement. On 10 November the Air Force officially reassigned it as a ground instructional airframe and transferred its ownership from AF Materiel Command to Air Combat Command. On 14 November 782 made its final flight when Dave Cooley, Chief Lockheed Martin F-117 Pilot, delivered it to Holloman AFB. During its distinctive flying career it had logged 1,535.5 hours and 1,205 flights.

Before retiring, 'Scorpion 3' was the oldest flying F-117 in the fleet with tail numbers 780 and 781 having already been retired. This leaves 'Scorpions 4' (783) and '5' (784) as the surviving FSD aircraft still flying with the 410th Flight Test Squadron. They are based with Lockheed Martin at Palmdale, California, but are assigned to the 412th Test Wing at Edwards AFB.

Terry Panopalis

During the Edwards show the resident 412th TW put up this fine Lockheed Martin formation comprising YF-117 782, two F-16Ds (87-392 and 92-455) and F-22A (91-009).

HLR requirement

Sikorsky CH-53K

On 5 January 2006 the US Naval Air Systems Command (NAVAIR) announced the assignment of the mission design series (MDS) designation CH-53K to the US Marine Corps' new heavy-lift helicopter. Built by Sikorsky Aircraft, and considered a near-term replacement for the CH-53E Super Stallion, the CH-53K will be capable of delivering a 13.5-ton load consisting of two up-armoured High Mobility Multipurpose Wheeled Vehicles (HMMWVs), one armoured vehicle, or three 9,000-lb (4082-kg) sustainment loads to three separate landing zones at a range of 100 nm (185 km).

Research, development, test and evaluation (RDT&E) efforts associated with the new helicopter began when Sikorsky was awarded a sole-source, $34 million contract, in support of the so-called CH-53X Heavy Lift Replacement (HLR) programme, on 23 December 2004. During August 2005 Sikorsky received an additional $43.3 million contract to continue with requirements definition, engineering trade studies and to proceed with risk reduction efforts.

Heavy Lift Replacement (HLR)

The HLR's history can be traced to late 2000 when the USMC announced plans to upgrade the CH-53E heavy-lift helicopter fleet. Introduced to the fleet in 1980, the three-engined CH-53E Super Stallion replaced the earlier twin-engined CH-53A/D Sea Stallion. Since then, however, the aircraft has suffered from degraded performance, fatigue, interoperability, maintainability and other operational issues. Additionally, the CH-53E costs approximately $17,500 per hour to operate and

requires 44.1 maintenance man-hours for each flight hour.

Although current plans call for the CH-53E to remain in service until at least 2025, the helicopters will begin reaching their aircraft's 6,150-hour design service life in the 2011/2012 timeframe, at a rate of 15 aircraft per year. Recent attrition has also caused NAVAIR to reclaim three former US Navy CH-53Es from storage at the Aerospace Maintenance and Regeneration Center (AMARC) at Davis Monthan AFB, Arizona. Once upgraded and brought up to the current configuration by Naval Air Depot Cherry Point, North Carolina, the helicopters will enter service with the Marines.

Then referred to as the CH-53X, the HLR project originally called for a remanufacture and service life extension programme (SLEP) for 111 of the service's 165 Super Stallions. The SLEP would have extended the aircraft's service life through 2025 at a cost of approximately $21 million each. The service expected to award a development contract during 2004 with full-rate production beginning in Fiscal Year 2011.

Consideration was also given to joining the US Army's Joint Heavy Lift (JHL) programme, which will develop a replacement for the CH-47 Chinook. However, the requirements for that project vastly exceed those of the Marines. Additionally, the JHL will not be ready for service before 2025 and will likely be too large to operate from the Navy's current or planned amphibious vessels. The service has, however, expressed interest in pursuing the JHL as a replacement for its C-130s.

An Operational Requirements Document (ORD) was completed in support of the HLR along with an analysis of alternatives (AOA) that evaluated seven existing aircraft platforms. As a result, engineers determined that only an enhanced CH-53 would meet its performance and

survivability requirements, operating and support costs, and the operational capability dates. The AOA also evaluated four alternative CH-53E designs and resulted in the decision to proceed with a design that met range and payload requirements and minimised the effects on service capability dates, inventory, support costs and risk. Completed in September 2003, the AOA determined that construction of new airframes was more cost-effective than upgrading the existing fleet of CH-53Es. Although engineers initially estimated that the remanufactured CH-53X would be about one-fifth the cost of a new replacement helicopter, experiences with the USMC's UH-1 and the US Army's CH-47 remanufacturing programmes have shown that the costs for those rebuilds are nearly equal to building a new airframe. In fact, the decision to build new UH-1Ys rather than remanufacture UH-1Ns is the result of a 1 per cent differential between the two approaches. Additionally, the remanufacture programmes require that a 'pipeline' of donor airframes be created and this would have had a negative impact on the service's ability to meet operational requirements. The ORD was subsequently approved by the Joint Requirements Oversight Council (JROC) in December 2003. In March 2004 the USMC announced plans to go ahead with the purchase of 154 new Heavy Lift Replacement (HLR) helicopters.

Development and production

The Defense Acquisition Board (DAB) gave its approval for HLR to enter the System Development and Demonstration (SDD) phase on 22 December 2005, and NAVAIR subsequently

The three-engined CH-53E has been in USMC service since June 1981. One of the roles it is used for is the recovery of downed aircraft. Here a Super Stallion prepares to lift an Iraqi air force Chengdu F-7 during aircraft recovery practice at Al Asad.

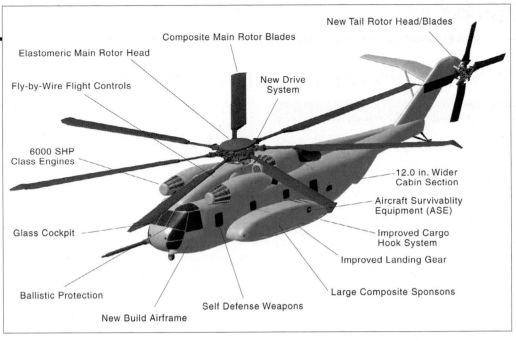

This Sikorsky impression summarises the key features of the CH-53K. Externally the main differences are the fuel-housing sponsons that replace the current external tank installation. The rotor blades are much wider and more efficient.

Elastomeric Main Rotor Head
Composite Main Rotor Blades
New Tail Rotor Head/Blades
Fly-by-Wire Flight Controls
New Drive System
6000 SHP Class Engines
12.0 in. Wider Cabin Section
Aircraft Survivablity Equipment (ASE)
Glass Cockpit
Improved Cargo Hook System
Improved Landing Gear
Large Composite Sponsons
Ballistic Protection
Self Defense Weapons
New Build Airframe

awarded Sikorsky a $8.4 million cost-plus, fixed-fee contract on 3 January 2006. A full SDD contract, valued at $2.9 billion, will likely be signed by March 2006. Sikorsky will continue its risk reduction and sub-system selection efforts as part of the interim SDD. Systems integration and research, development, test & evaluation (RDT&E) efforts will run through 2015 and are expected to cost up to $4.4 billion.

Preliminary (PDR) and critical design reviews (CDR) are respectively planned for 2006 and 2008, and the construction of five prototype YCH-53Ks will follow. Five SDD airframes will include a single ground test vehicle (GTV) and four flying engineering development models (EDM). The GTV will be delivered to Sikorsky's flight test facility in West Palm Beach, Florida, in late 2010. The first EDM example will follow in mid-2011 and will make its initial flight in late 2011. Initial flight testing of the EDM aircraft will be carried out by the HLR Integrated Test Team (ITT) at the Florida facility. However, the aircraft will be ferried to NAS Patuxent River, Maryland, in 2012/13 where the ITT will carry out the majority of development testing.

GTV	Propulsion system development & qualification, systems integration, climatic tests, live-fire tests
EDM 1	Structural/dynamics
EDM 2	Handling qualities, performance, propulsion
EDM 3	Avionics, electrical, cargo, and environmental
EDM 4	Electromagnetic environmental effects (E3), maintainability

The first of three Low Rate Initial Production (LRIP) lots will be ordered in Fiscal Year (FY) 2013 and they will respectively include 6, 9 and 14 aircraft. The initial production aircraft will be delivered in 2014 and fleet introduction will follow in 2015. The CH-53K will also achieve initial operational capability (IOC) in 2015, when four aircraft and combat-ready crews are declared to be 'ready for deployment'. Full operational capability (FOC) will follow by FY 2021. Each aircraft is expected to cost an average of $56 million, with the total programme procurement cost reaching $14.4 billion.

Whereas the CH-53E is capable of operating at a maximum gross weight of 73,000 lb (33112 kg), the CH-53K's gross weight is increased to 84,700 lb (38420 kg). Visibility for the pilots will be improved through the incorporation of larger windscreens and side panels. Additionally, its fuselage and cabin will be 12 in (30.5 cm) wider than the earlier model, allowing it to carry C-130-size 463L series pallets. Although troop seating is reduced from 55 in the CH-53E to just 30 aboard the CH-53K, the passengers will be provided with newly designed crashworthy seats. Cargo movement will also be aided by an internal cargo rail locking system and an improved cargo hook system.

Larger external sponsons will increase the helicopter's internal fuel load, alleviating the need for external fuel tanks. Although not as large as those found on the US Navy's MH-53E, the sponsons will be constructed from composite materials and will be longer and taller than those of the CH-53E. Its deck footprint will remain unchanged, despite the installation of improved landing gear. The CH-53K combat radius will be double that of the CH-53E and its high/hot performance will enable it to carry 27,000 lb (12247 kg) of cargo 110 nm (204 km), to an altitude of 3,000 ft (914 m) at an ambient temperature of 91.5°F (33°C).

The CH-53E's T64-GE-416 turboshaft engines will be replaced by new engines in the 6,000 shp (4476-kW) class. Rolls-Royce's AE1107C turboshaft engine, which already powers the USMC's MV-22B Osprey, is certainly a candidate. Rated at 6,150 shp (4588 kW) each, it would provide a total increase of 4,210 shp (3,140 kW) over the T64. A new split-torque gearbox, an elastomeric rotor head and high-efficiency composite rotor blades featuring an advanced airfoil and swept tips will also be incorporated. Although the diameters of the main and tail rotors will likely remain unchanged, the chord of both will be increased. The main rotor blades will be based on those used by Sikorsky's commercial S-92 helicopter.

It will also feature a 'fly-by-wire' flight control system and a so-called Joint Interoperable 'Glass' cockpit. The latter will provide the helicopter with an electronic flight instrument system (EFIS) that could be based on that of either the MV-22B, UH-1Y or CH-60S/MH-60R helicopter. Whereas the CH-53E was intended to operate in benign environments, the CH-53K is designed to go in harm's way and will be equipped with defensive countermeasures, ballistic protection and weapons including the GAU-21 0.5-in (12.7-mm) machine-gun. The CH-53K will provide the fleet with improvements in range and payload performance, cargo-handling, turn-around times and survivability. Additionally, the advanced technologies incorporated in the helicopter will result in a major reduction in life cycle costs over the CH-53E, while improving reliability, maintainability and interoperability.

Tom Kaminski

Production plans

Year	Aircraft
FY13	6
FY14	9
FY15	14
FY16	21
FY17	24
FY18	24
FY19	24
FY20	24
FY21	10
Total	**156**

Specifications

	CH-53E	CH-53K
Max. gross weight	73,000 lb (33112 kg)	84,700 lb (38420 kg)
Max. external cargo (to 110 nm)	12,100 lb (5488 kg)	27,000 lb (12247 kg)
Mission radius (unrefuelled, no load)	210 nm (389 km)	200 nm (370 km)
Max. passengers	55	30
Max. airspeed	150 kt (278 km/h)	150 kt (278 km/h)

The CH-53K will have three underfuselage cargo hooks. The fore and aft hooks will be rated at 25,000 lb (11340 kg) each while the central hook will be able to carry up to 36,000 lb (16330 kg).

New utility helicopter for the US Army

LUH competition

During February 2004 the US Army Aviation and Missile Command (AMCOM) announced its intention to purchase a new Light Utility Helicopter (LUH), using funds that had been earmarked for the cancelled RAH-66 Comanche helicopter programme. A draft request for proposals was released to industry on 4 November 2004 and the formal RFP followed in July 2005. As part of its evaluation, the army tested each of the competing helicopters during a Source Selection Performance Demonstration (SSPD) that was conducted by the Aviation Technical Test Center (ATTC) at Cairns Army Airfield at Fort Rucker, Alabama, during February 2006.

The army intends to select a winner in April 2006, and will purchase 26 helicopters under low-rate initial production (LRIP) contracts. The Fiscal Year 2006 appropriations have secured funding for the initial 12 examples. The first LUH will be delivered in September 2006 and the helicopter will achieve initial operational capability (IOC) in October 2008. The final examples will be delivered in 2013. Besides aircraft production, the $1.5 billion project includes the delivery of MEDEVAC and hoist kits and contractor logistic support (CLS). In addition, the winning contractor will be responsible for all training associated with the

EADS has used EC145/Bk-117C2) N145UH as a UH-145 demonstrator, seen here displaying the clamshell cabin doors that facilitate loading.

LUH, including aviators, crewmembers and maintainers.

LUHs will be assigned to the Army National Guard and specific Active Component (AC) Army units, replacing the service's remaining UH-1H and OH-58A/C helicopters. Known as Table of Organizations and Equipment (TOE) and Table of Distribution and Allowances (TDA) units, the latter include both operational field units and non-deployable organisations. The helicopters will, respectively, be divided between the active army and ARNG, with the former acquiring 118 examples and the latter 204. Within the ARNG, the LUH will conduct light general support, civil search and rescue (SAR), personnel recovery, air ambulance medical evacuation (MEDEVAC), casualty evacuation (CASEVAC), limited civil command and control operations in support of Homeland Security (HLS), and counter-drug operations. Although deployable, the aircraft will only perform these functions in permissive, non-hostile, non-combat operational environments.

Those aircraft assigned to the TOE units are tasked with light general support (GS) missions that include the time-sensitive transportation of equipment, supplies, documents and/or personnel. Among the roles assigned to the active duty TDA units, the LUH will serve as observer/controller aircraft at Combat Training Centers (CTC). They will also conduct force protection and installation security over sensitive areas, including test sites and ranges, and serve as chase/instrumentation aircraft for technical and operational testing.

With the capability to lift 15 passengers or six stretchers plus attendants, the US139 is the largest of the four competitors, and along with the Bell 412EP represents the 'large' approach to the LUH competition.

Chief among the LUH specifications was the requirement that the helicopter must be a commercial off-the-shelf (COTS), non-developmental item (NDI) and must already hold a Federal Aviation Administration (FAA) Type Certificate. Although provided with two crew stations the LUH must be operable by a single pilot under visual flight regulations (VFR), and instrumentation must be compatible with night vision goggles (NVG). It will be capable of self-deploying to any location within the CONUS but will also be transportable by air aboard C-17A and C-5A/B-class airlfiters.

In the MEDEVAC role the LUH cabin will provide accommodations for at least two standard NATO litters, a medical attendant with equipment and at least one additional passenger. For non-medical missions it will be capable of carrying at least six passengers. It will also be provided with a 600-lb (272-kg) capacity rescue hoist and will support firefighting operations when equipped with a 'bambi-bucket'. Its internal load capacity will be a minimum of 1,500 lb (680 kg) and an external load of up to 2,200 lb (998 kg). The helicopter will be capable of a cruise speed of 125 kt (232 km/h) and its operational range must be at least 217 nm (402 km) when operating in a high/hot environment. The aircraft will have a minimum endurance of 2.8 hours without carrying auxiliary fuel or using a forward arming and refuelling point (FARP).

Under the current fielding schedule, the last aircraft replaced by the LUH will be the OH-58A/Cs assigned to the Army National Guard's Reconnaissance and Interdiction Detachments (RAID)/Security and Support (S&S) Battalions. Accordingly, the LUH will provide the necessary provisions for up to 200 lb (91 kg) of mission equipment including forward-looking infrared (FLIR), video downlink, searchlight and communications equipment, which will be integrated later.

Specifications

	US139	B412EP	UH-145	MD902
Length overall	54.7 ft (16.66 m)	52.1 ft (15.89 m)	42.7 ft (13.01 m)	38.8 ft (11.83 m)
Fuselage length	44.4 ft (13.52m)	43.3 ft (12.91 m)	33.4 ft (10.18 m)	32.3 ft (9.85 m)
Height overall	16.2 ft (4.95m)	14.9 ft (4.54 m)	11.3 ft (3.44 m)	12 ft (3.66 m)
Fuselage width	10 ft (3.04m)	9.4 ft (2.86 m)	5.7 ft (1.74 m)	9.3 ft (2.83 m)
Main rotor diameter	45.3 ft (13.80 m)	46.0 ft (14.02 m)	36.1 ft (11.03 m)	33.8 ft (10.30 m)
Tail rotor diameter	8.83 ft (2.70 m)	8.5 ft (2.58 m)	6.4 ft (1.95 m)	(NOTAR system)
Maximum take-off weight	14,110 lb (6,400 kg)	11,900 lb (5,398 kg)	7,903 lb (3,585 kg)	6,900 lb (3,130 kg)
Useful load	6,019 lb (2,730 kg)	4,803 lb (2,180 kg)	3,953 lb (1,793 kg)	3,125 lb (1,417 kg)
External load	n/a	4,500 lb (2,041 kg)	3,307 lb (1,500 kg)	3,000 lb (1,361 kg)
Powerplant	2 P&WC PT6C-67C	2 P&WC PT6T-3D	2 Turbomeca Ariel 1E2	2 P&W PW207E
Maximum power rating	1,679 shp (1,252 kW)	1,800 shp (1,342 kW)	1,456 shp (1,100 kW)	1,420 shp (1,059 kW)
Maximum speed	167 kt (309 km/h)	140 kt (259 km/h)	145 kt (269 km/h)	140 kt (259 km/h)
Cruise speed	157 kt (291 km/h)	128 kt (237 km/h)	131 kt (243 km/h)	134 kt (248 km/h)
Service ceiling	19,460 ft (5,931 m)	n/a	18,000 ft (5,486 m)	17,600 ft (5,364 m)
Hover ceiling (IGE)	15,660 ft (4,773 m)	17,400 ft (4,304 m)	11,300 ft (3,444 m)	10,800 ft (3,292 m)
Hover ceiling (OGE)	12,014 ft (3,662 m)	13,800 ft (4,206 m)	9,000 ft (2,743 m)	8,800 ft (2,682 m)
Range	307 nm (568 km)	423 nm (784 km)	370 nm (685 km)	225 nm (417 km)
Crew	2	2	2	2
Accommodation	15 passengers or 6 stretchers and 2 attendants	13 passengers or 6 stretchers and 2 attendants	8 passengers or 2 stretchers and 2 attendants	7 passengers 2 stretchers and 2 attendants
Cost	$8 million+	$6.5-7 million	$4.5 million	$3.5 million

LUH notional fielding schedule

Active Army

Location	FY	Qty
Germany (two locations)	FY06/FY09	18
Fort Rucker, Alabama	FY07	6
Fort Irwin, California	FY08/FY09, FY13	33
Fort Polk, Louisiana	FY08/FY09, FY13	24
Fort Belvoir, Virginia (NVL)	FY09	2
Fort Eustis, Virginia (AATD)	FY09	2
Fort Hood, Texas	FY09	2
Fort Rucker, Alabama (ATTC)	FY09	4
Aberdeen Proving Grounds, Md.	FY09	4
USA Kwajalein Test Center, HI	FY09	4
NAEC Lakehurst, N.J. (RDEC)	FY09	2
White Sands Missile Range, N.M.	FY09	6
USMA (2nd AVN DET), New York	FY09	2
Yuma Proving Grounds, Arizona	FY09	5
Total		**118**

Army National Guard

Unit	FY	Qty
MEDEVAC	FY07/08	60
S&S Battalions (6)*	FY09/FY13	144
Total		**204**

Six Aviation Security & Support Battalions will each be equipped with 24 aircraft assigned to three companies. Aircraft will be assigned to units in 44 states, the District of Columbia and Puerto Rico. Six states will operate the aircraft from multiple locations.

The contractors that responded to the LUH RFP included AgustaWestland with the US139, Bell Helicopter with its Model 412EP, EADS North America with the UH-145, and MD Helicopters with the MD902 Explorer.

AgustaWestland

AgustaWestland, Inc. (AWI) was initially expected to offer its A109 for the LUH competition. In a surprise move, AWI, which is headquartered in Reston, Virginia, teamed with L-3 Communications Integrated Systems (L-3/IS) in Waco, Texas, to propose a variant of its AB139. Although Bell/Agusta Aerospace developed the AB139, AgustaWestland and Bell Helicopter recently revised their joint venture partnership leaving the former to market the AB139. The so-called US139 is the largest and arguably the most capable of the four entrants. However, it is also the most expensive at around $8 million per copy. It is powered by a pair of Pratt & Whitney Canada PT6C-67C turboshaft engines and is equipped with five-blade main and four-blade tail rotors. The helicopter is also equipped with retractable landing gear and the cockpit features Honeywell's Primus Epic integrated avionics system. Under the terms of the teaming agreement AWI will be responsible for producing the US139 airframes, whereas final assembly, integration and training will be carried out by principal partner L-3 at its facility in Waco. The latter contractor will also be responsible for the contractor logistic support (CLS) functions.

Bell Helicopter

Bell Helicopter, which is headquartered in Fort Worth, Texas, had originally planned to offer a modernised UH-1H known as the Bell 210. In fact, the aircraft had received its FAA certification when the army revised its LUH requirements. The decision to pursue a twin-engined aircraft certified for instrument flight, however, effectively ruled out the 210. Bell subsequently chose to offer the model 412EP, which is built at its facility in Mirabel, Quebec. Specific components are produced at the company's Fort Worth, Texas, and Piney Flats, Tennessee, facilities. Bell also intends to conduct training in Fort Worth if the 412 is selected.

The 412EP is powered by two Pratt & Whitney Canada PT6C-3D turboshaft engines and equipped with four-bladed main and tail rotors, and costs approximately $6.5-7 million each. The similar model 412HP is already in service with the Canadian Air Force, which flies nearly 100 examples under the designation CH-146. Operated in support of the Canadian Army, the CH-146 already carries out duties similar to those planned for the LUH.

EADS North America

Teamed with its own American Eurocopter business unit, as well as Sikorsky Aircraft, Westwind Technologies and CAE USA, EADS has proposed a slightly modified version of Eurocopter's EC145 twin-engined commercial helicopter for the LUH. The helicopter is powered by a pair of Turbomeca Arriel 1E2 turboshaft engines, and is equipped with a four-blade hingeless rotor system and composite rotor blades. In addition to sliding doors on each side of the cabin, the UH-145 offers large clamshell doors at the rear of the cabin, which allow for easy loading and unloading. Each helicopter costs about $4.5 million. Although EADS, which will serve as the prime contractor, plans to build the initial 50 examples of the so-called UH-145 in Germany, the aircraft will be completed at the American Eurocopter facility in Columbus, Mississippi. The remainder of the aircraft will all be produced by the Columbus facility. Sikorsky will be responsible for providing the CLS and Westwind Technologies, which manages a logistics support and modification facility at Redstone Arsenal on behalf of the Army's Program Executive Office (PEO), Aviation, will provide engineering and programme management support. CAE USA, which is located in Tampa, Florida, will supply the UH-145 cockpit procedural trainers that will support pilot training and familiarisation. During late January EADS announced that production of two EC145s had begun, in Germany, and those aircraft would be delivered to the US Army should the company win the LUH competition.

MD Helicopter

The MD Explorer (MD902) is the smallest and least expensive entrant in the LUH competition at around $3.5 million per aircraft. It also offers the lowest direct operating costs and is the only completely 'American' entrant. MD Helicopters Inc. (MDHI) originally entered the competition teamed with Lockheed Martin but the partnership was later dissolved. MDHI subsequently teamed with DynCorp International, Aviation Systems of Northwest Florida (ASI) and GENCO Infrastructure Solutions. Whereas MDHI will build the aircraft and serve as the prime contractor, DynCorp will provide the CLS. GENCO will provide the training devices and will support CLS operations. Powered by two Pratt & Whitney P207E turboshaft engines, the MD Explorer is equipped with a five-blade main rotor and features the NOTAR (no-tail-rotor) system for anti-torque and directional control. In addition to sliding main cabin doors, the aircraft is equipped with large rear clamshell doors for loading and unloading cargo.

Tom Kaminski

Below: N912LH is MDH's Explorer LUH demonstrator, seen here during a test flight near the company's main facility at Mesa, Arizona.

Right: Having ditched the single-engined Model 210, Bell is offering the 412EP, using this aircraft (N44438) to display the type's capabilities.

JCA – Army/Air Force airlifter

Joint Cargo Aircraft

On 1 February 2006 the US Army and Air Force Chiefs of Staff signed a memorandum of understanding (MoU) through which the two services consolidated their respective Future and Light Cargo Aircraft (FCA/LCA) efforts under a single programme. The combined project was subsequently renamed the Joint Cargo Aircraft (JCA) on 1 March 2006. The joint MoU does not require the selection of a single aircraft but one airframe will likely be adapted to fulfill both missions.

Although the USAF had not then formalised its LCA requirements, the Army released a draft request for proposals (RFP) for its FCA in August 2005 and was expected to issue the formal request in late 2005. The service had planned to select a winner in June 2006. However, the project was put on hold when the Department of Defense directed that the FCA and LCA be jointly developed. It now appears that the request for the joint airlift programme will be released in March 2006 and a winner will be selected in December. The joint project will reportedly include at least 140 aircraft, comprising 70 each for the Army and the USAF, although the latter could receive as many as 150 examples.

Future Cargo Aircraft (FCA)

Originally known as the CX-X, the FCA called for an intra-theatre support aircraft capable of delivering priority cargo for the US Army. The

Logistics Support Area Anaconda, located at Balad Air Base in northern Iraq, was home to numerous Army National Guard C-23B Sherpas that were responsible for moving materials, and personnel throughout Iraq.

service initially planned to purchase 25 examples of the CX-X using funding reprogrammed in the aftermath of the RAH-66 Comanche helicopter's cancellation, but later announced plans to order 45 aircraft between 2007 to 2011. Intended as a replacement for the US Army National Guard's fleet of 43 C-23B/B+ support aircraft, it would begin replacing the Sherpa in 2007. The programme would run through 2025 and the FCA would later replace the C-26 and selected C-12s. Most recently, the Army planned to acquire 33 FCAs at a cost of $1.3 billion, purchasing three examples in FY 2007 followed by four, seven, eight and 11 examples annually through 2011. The army's budget for 2006 earmarked $4.9 million for the programme and $109 million was requested for FY 2007.

The requirement for the FCA was highlighted by the recent inability of the C-23 to support the Army's needs during Operation Enduring Freedom in Afghanistan, which forced the service to use the CH-47D helicopter and contracted fixed-wing airlifters to carry out intra-theatre missions. Although the helicopter is capable of carrying up to 18,000 lb (8165 kg) it has a relatively short range. Conversely, the Sherpa has a longer range but can only carry around 7,000 lb (3175 kg) of cargo and it is not pressurised, which limited its ability to fly at the higher altitudes required in Afghanistan. The JCA will provide the service with a single aircraft that combines the Chinook's cargo capacity with a longer range.

The airlifter will be tasked with performing logistical resupply, casualty evacuation (CASEVAC), troop movement and air drop operations, as well as humanitarian and homeland security missions. One key mission will be in support of the Army's new Brigade Combat Teams (BCT) and involves delivering mission-critical and time-sensitive supplies and key personnel from intermediate staging areas directly to the BCTs. Accordingly, the FCA must be capable of operating in high-threat environments, and in remote and austere locations. The pressurised, twin-engined aircraft will have a self-deployment range of 2,100 nm (3889 km) and a combat radius of 600 nm (1111 km). It will be capable of carrying NATO 463L-series pallets, and performing short take-offs and landings on 2,000-ft (610-m) unimproved runways.

Light Cargo Aircraft

In response to its own experiences in Afghanistan and Iraq, the USAF unveiled plans for a Light Cargo Aircraft (LCA) during 2005. The LCA would be capable of carrying two pallets or 25 to 30 troops operating from 2,000- to 2,500-ft (610- to 762-m) airstrips. The aircraft shared several missions with the army's FCA and was expected to support Air National Guard Homeland Security (HLS) and humanitarian efforts that were highlighted following the recent hurricanes which struck along the Gulf coast. Air National Guard officials hoped to assign a fleet of eight to 12 aircraft to each of the 10 Federal Emergency Management Agency (FEMA) regions. The USAF initially planned to purchase 24 aircraft and had requested $15.8 million in its FY2007 budget to begin the LCA competition. Interestingly, the Air Force Special Operational Command (AFSOC) recently placed six Pilatus PC-12 light transports in service under the designation U-28A. The aircraft, which fulfilled a more immediate need, are primarily intended to provide intra-theatre airlift for special operations forces.

Joint programme

Two teams that have already invested heavily in the JCA competition include Global Military Aircraft Systems (GMAS), and the so-called Team JCA. Formed as a 50/50 joint venture between L3 Communications Integrated Systems (L3 IS) and Alenia Aeronautica subsidiary Alenia North America Inc., GMAS is marketing the C-27J. L3IS, then known as Chrysler Technologies Airborne Systems, previously upgraded 10 Alenia G.222-710 airlifters and delivered them to the USAF under the designation C-27A. The aircraft supported US Southern Command until retired in 1999, and several examples are currently flown by the US Department of State Air Wing. Team JCA, which includes Raytheon Company's Space and Airborne Systems business unit and EADS-CASA North America, may offer the C-295, the CN-235 or a mixed fleet including the C-295 and CN-235. Although the three aircraft are generally comparable, the C-27J can fly faster, farther and higher, and carry more cargo than the C-295 or CN-235, which are less expensive to purchase, operate and maintain.

The joint project could result in additional bidders stepping forward and Boeing is one manufacturer that recently expressed interest in the project. The contractor is considering a proposal

that would include a variant of the 70-seat, turbofan-powered Antonov An-72 airlifter. The Ukrainian aircraft would likely require a considerable design effort in order to equip it with western engines and avionics systems.

C-27J

Developed by Alenia Aeronautica and Lockheed Martin, the C-27J is based on the earlier G.222, which first flew in 1975. It is powered by the same Rolls-Royce engines that equip Lockheed Martin's C-130J and flew for the first time in September 1999. Capable of carrying a 13,227-lb (6000-kg) payload a distance of 2,300 nm (4260 km), the C-27J has a top speed of 325 kt (610 km/h) and a maximum altitude of 30,000 ft (9144 m). Its cargo compartment can carry three HCU-6 88 x 108-in (223 x 274-cm) and one HCU-12 54 x 88-in (137 x 223-cm) pallets, or seven HCU-12 463L-series cargo pallets. Its floor is stressed for a maximum loading of 4900 kg/m² (1,004 lb/sq ft), exceeding that of the C-130 and enabling the aircraft to carry light armored vehicles. The C-27J is already in service with the Italian and Hellenic Air Forces and was recently selected by Bulgaria. GMAS currently plans to offer the aircraft to the Canadian Air Force as a rescue aircraft and to the US Special Operations Command as a combat rescue tanker.

C-295

Designed by EADS-CASA division in Spain, the C-295 – which first flew in 1998 – is currently in service with the air forces of Jordan, Spain and Poland and has been selected by Brazil and most recently Portugal. It is a stretched version of the earlier CN-235, which was developed by Airtech (a joint venture between CASA and IPTN in Indonesia) and first flew in 1983. The aircraft's 41-ft (12-m) long cabin is capable of carrying up to a 20,400-lb (9253-kg) payload that would typically include three vehicles, three jet engines or five 88 x 108-in (223 x 274-cm) 463L cargo pallets. The aircraft has a top speed of 260 kt (482 km/h), can operate at an altitude of 25,000 ft (7620 m) and has a range of 2,300 nm (4260 km) when carrying a 10,000-lb (4536-kg) payload.

CN-235

Originally developed by CASA/IPTN, through the Airtech joint venture, the CN-235 first flew in November 1983. Currently built by EADS-CASA

in Spain and by Dirgantara Indonesia, the military transport variant is in service with numerous air forces worldwide. The transport has a 31-ft (9.5-m) long cabin and is capable of carrying four 463L pallets or up to a 13,100-lb (5942-kg) payload. The CN-235 has a range of 1,230 nm (2278 km) when carrying a 10,000-lb (4536-kg) payload and a top speed of 245 kt (454 km/h). Like the C-295, it has a service ceiling of 25,000 ft (7620 m).

Raytheon is the prime contractor for the Team JCA effort and is responsible for program management, mission systems integration and

Alenia is no longer working with Lockheed Martin on the C-27J, and is now teamed with L3/IS. The C-27J uses the engines, cockpit and other systems from the C-130J. This is the third prototype, converted from a G.222TCM.

mission support solutions. EADS CASA North America will assemble and deliver the aircraft at its newly opened facility in Mobile, Alabama.

Tom Kaminski

Representing the Team JCA proposal is this formation of the first production-representative C-295 in the foreground, accompanied by the CN-235-300 demonstrator.

Specifications

	C-23B+	C-27J	C-295	CN-235
Fuselage length	58.04 ft (17.69 m)	74.48 ft (22.70 m)	80.33 ft (24.48 m)	70.12 ft (21.37 m)
Height	16.42 ft (5.00 m)	31.82 ft (9.70 m)	28.42 ft (8.66 m)	26.83 ft (8.18 m)
Wingspan	74.83 ft (22.81 m)	94.16 ft (28.70 m)	84.66 ft (25.80 m)	84.67 ft (25.81 m)
Cabin length	29.83 ft (9.09 m)	35.50 ft (11.43 m) including ramp	41.66 ft (12.70 m) w/o ramp	31.67 ft (9.65 m)
Cabin width	6.22 ft (1.90 m)	8.04 ft (2.45 m)	8.83 ft (2.69 m)	8.67 ft (2.64 m)
Cabin height	6.50 ft (1.98 m)	8.53 ft (2.6 m)	6.25 ft (1.91 m)	6.25 ft (1.91 m)
Maximum take-off weight	25,600 lb (11,612 kg)	70,107 lb (31,800 kg)	51,150 lb (23,201 kg)	36,830 lb (16706 kg)
Maximum payload	7,280 lb (3,302 kg)	25,353 lb (11,500 kg)	20,400 lb (9,253 kg)	13,100 lb (5942 kg)
Fuel	679 gal (2,570 l)	3,255 gal (12,320 l)	2,034 gal (7,700 l)	2,034 gal (7,700 l)
Powerplant	2 x 1,424-shp (1,062-kW) P&WC PT6A-65AR	2 x 4,637-shp (1,252-kW) Rolls-Royce AE 2100-D2	2 x 2,645-shp (1,972-kW) P&W PW127G turboprop	2 x 1,750-shp (1305-kW) GE CT7-9C3 turboprop
Propeller	5-blade Hartzell HC-B5MP-3C	6-blade Dowty R-391	6-blade Hamilton Sundstrand 568F-5	4-blade Hamilton Sundstrand 14RF-37
Maximum speed	240 kt (444 km/h)	325 kt (610 km/h)	260 kt (482 km/h)	245 kt (454 km/h)
Ceiling	13,950 ft (4,252 m)	30,000 ft (9,144 m)	25,000 ft (7,620 m)	25,000 ft (7,620 m)
Range	669 nm (1,239 km) w/5,000-lb (2268-kg) payload	2,300 nm (4,260 km) w/13,227-lb (6000-kg) payload	2,300 nm (4,260 km) w/10,000-lb (4536-kg) payload	1,230 nm (2278 km) w/10,000-lb (4536-kg) payload
Ferry range	869 nm (1,609 km)	3,200 nm (5,926 km)	3,040 nm (5,630 km)	2,730 nm (5056 km)
Take-off roll	2,630 ft (802 m)	1,903 ft (580 m)	2,200 ft (671 m)	1,325 ft (404 m)
Landing roll	1,920 ft (585 m)	1,115 ft (340 m)	1,050 ft (320 m)	1,240 ft (378 m)
Crew	3	2	2	2
Accommodation	27 troops or 30 passengers or 15 stretchers and 3 attendants	68 troops or 15 passengers or 36 stretchers and 6 attendants	71 troops or 51 passengers or 24 stretchers and 4 attendants	51 troops or 24 stretchers and 4 attendants

Saab UAV development

SHARC and FILUR

From the 1990s it became apparent that unmanned air vehicles (UAVs) would play an increasingly important part in military operations. In common with other forward-thinking nations, Sweden has been involved in UAV research for some years, and has now flown two key demonstrators – SHARC and FILUR – as Saab embarks on a major contribution to the European Neuron UCAV programme.

Saab's work on UAVs began formally in early 1998 as part of the Nationellt Flygteknisk Forsknings-Program (NFFP – National Aeronautics Research Programme) that co-ordinates the research and development efforts of industry, research establishments and universities. Project NFFP 272 involved Saab, Ericsson and other agencies, and began in April to June 1998 with the definition of nine potential configurations to represent various UAV design and construction concepts. Subsequent evaluations led to the selection of one configuration, which was christened SHARC (Swedish Highly Advanced Research Configuration). This represented a low-observable, attack-capable UCAV (that would carry its non-stealthy weapons internally). In March 1999 a low-speed model was tested in the wind tunnel at the FFA (Flygtekniska Försöksanstalten), a part of the FOI (Defence Research Agency), in Stockholm. The trials included tests in the T1500 transonic tunnel, and simulated weapons release from an internal bay.

Construction of the SHARC demonstrators got under way in 2001. The SHARC vehicle is not stealthy, but its shape represents that of a potential stealthy UCAV. The vehicle was developed rapidly so that a Swedish UAV could be put in the air to provide 'hands-on' knowledge of representative UAV operations and design, as well as to help engineers to examine and solve the problems of UAV certification issues. Swedish forces already have some UAV experience in the shape of the SAGEM Sperwer tactical UAV, known locally as the Ugglan (owl).

The SHARC configuration first flew (as the 'Baby SHARC' model) on 11 February 2002 at an undisclosed location. Four SHARC demonstrators were built. The first two radio-controlled aircraft (BT-001 and BT-002) had retractable undercarriages and were used for aerodynamic tests and later for UAV operator training. They had a pitot tube protruding from the aircraft's spine, and a raised intake for the small Olympus 210N jet engine. The second two machines (BS-001 and BS-002) differed by having fixed undercarriage, long air data booms projecting from the nose, a GPS antenna above the nose and a buried air intake for the engine.

Initially the aircraft was hand-flown by radio control, but once the aerodynamic and flight-handling properties had been verified, a greater degree of autonomy was introduced as the tests progressed. During the second test campaign the SHARC flew autonomous mid-flight sectors. During the third campaign – with differential GPS

SHARC BS-002 flies over the forests near Vidsel. This vehicle was produced to demonstrate fully autonomous operations in a UCAV-style vehicle, a feat it successfully achieved in August 2004 to bring the SHARC programme to a successful conclusion.

(DGPS) and a radar altimeter installed and integrated with the flight control system – the SHARC performed its first fully automatic flight. On 25 August 2004 a SHARC took off from Vidsel in northern Sweden, performed a simulated mission and landed, all without human intervention. This first fully autonomous flight marked a major milestone in the Swedish UAV programme and paved the way for further developments.

Stealth demonstrator

Having proven that autonomous operations could be safely undertaken, the Saab team turned to work on a new vehicle that would test operational applications, principally in the area of stealth technology. Using stealth as part of an attack UAV design has obvious survivability implications, but it also drives costs down as there is theoretically no need to carry defences. Incidentally, Sweden's research agencies have been involved in stealth studies since the 1970s.

Saab's stealth demonstrator is named FILUR (officially Flying Innovative Low-observable Unmanned Research, although 'filur' is also Swedish slang for a 'sly dog'). Funding of the 80 million kronor programme mainly comes from the FMV (Försvarets Materielverk – Defence Material Administration). It is a sub-scale demonstrator of a larger, generic UCAV. As well as its stealthy flying-wing shape, it incorporates radar absorbent material and carbon-fibre reinforced plastic in its construction. The engine intake is mounted on the spine, well out of the way of prying radars, and the 2.26-kN (507-lb) thrust AMT Olympus 210N engine exhausts through a broad, flattened slit nozzle in the pointed rear of the fuselage. Overall dimensions are 2.2 m (7 ft 3 in) in length and 2.5 m (8 ft 2 in) in span. It weighs around 55 kg (121 lb). Control is effected by split surfaces on the trailing edge of the wings/rear fuselage.

FILUR uses the computer/flight control system already proven in the autonomous SHARC. To reduce the risks associated with flying the new aircraft, a small-scale aerodynamically representative model was built and flown in spring 2004. Known affectionately as 'Baby FILUR', the model was built in conjunction with the University of Linköping, with which Saab is closely associated through the joint LinkLab research facility.

Ground-based radar cross-section tests were performed with FILUR in late 2004 and showed considerable promise, as well as tallying well with predictions. On 10 October 2005 the FILUR (FLR-01) undertook its first flight at the Vidsel test airfield near the Arctic Circle. The successful test lasted for about 10 minutes, and provided sufficient data to allow engineers to assess flying characteristics. When flight tests resume in the early summer of 2006, it is expected that they will jump straight into the main research programme, which is aimed at testing stealth aspects. As well as

Above: The four SHARC airframes are divided between remotely-piloted test/training aircraft (BT- serials) and fully autonomous vehicles (BS- serials). Despite its looks, SHARC is not a stealthy vehicle. Top speed is in the order of 170 kt (315 km/h).

Below: FILUR has a very clean airframe, necessary to achieve low radar signatures. The engine intake is mounted on top of the aircraft, where it cannot be seen by ground-based radars, and leads to an S-shaped duct that further reduces radar returns.

Gripen is displayed with its two smaller stablemates. There is a considerable amount of technology cross-over between the manned Gripen fighter and the UCAV programme. The central computer of the Gripen was used as the basis of the avionics systems of both SHARC and FILUR. One concept that has received much attention is that of using a two-seat Gripen to control fleets of UCAVs during attack missions.

signature measurements associated with the vehicle itself, the FILUR will also fly against various Swedish defence systems to test their capabilities against low-observable aircraft.

For the first flight the air vehicle's undercarriage was fixed, but for the follow-on stealth tests it will be retractable. It was also hand-flown remotely for this first test, the 'pilot' being provided with video from a forward-facing TV camera. Subsequent flights will involve an incremental increase in autonomy, leading to fully autonomous flights as proven by SHARC. FILUR also has removable vertical fins attached. According to ground tests and computer predictions, the aircraft should be stable without them. However, for early test flights they were fitted as an insurance measure. As they are almost transparent to radar, the fins will not affect stealth measurements too much if they have to be retained.

Neuron

Data from the FILUR programme will be fed into the European Neuron unmanned combat air vehicle. Saab is an important partner in this Dassault-led project, with low-observable technol-

ogy (especially of the weapon bay doors and engine nozzle) as one of its areas of responsibility. Other Saab responsibilities include avionics management, fuel system and aerodynamic analysis/shaping. Neuron was launched in June 2003, but full Swedish participation was held up due to domestic political wrangling. Following a government report concerning potential options, the go-ahead was given on 20 December 2005. Of the total Swedish commitment of 750 million kronor, Saab is contributing 600 million. In return, the company received guarantees of a similar figure back from the government for investment in ongoing Gripen developments.

As well as Neuron, Saab has also been studying other classes of operational UAV. The TUAV (Tactical UAV) is a small, low-cost UAV for tactical reconnaissance and other short-range missions, able to carry a sensor payload of over

50 kg (110 lb). There is a Swedish requirement to replace the Ugglan in support of the Nordic battlegroup (forces from Estonia, Finland, Norway and Sweden) that is due to stand up in January 2008. Saab is talking to several potential partners, and development work is being aimed at reducing the 'footprint' of such UAVs in terms of support personnel and equipment. An RFP is expected in mid/late 2006.

The MALE (Medium-Altitude, Long Endurance) UAV is intended for long-range surveillance, attack and EW missions, and could be used for communications relay. Although some work has been done, this class is of lower priority for Saab, while the company continues to monitor multi-national EuroMALE developments. Saab is also heavily involved in UAV certification issues, for which it is preparing for Demo 07, a demonstration of a UAV operating in non-segregated airspace to be held in conjunction with civil agencies. This may involve an actual UAV or a manned surrogate – a decision to be made in spring 2006.

David Donald

FILUR gets air under its wheels for the first time at the RFN Vidsel test site. The Dayglo paint was added for conspicuity during early test flights. The first flight verified much of the handling of the vehicle, and when flying resumes in 2006 after a long winter lay-up the accent will be on tests covering both infra-red and radar signatures. Vidsel is scheduled to be one of the sites used for flight tests of the Neuron UCAV from 2011.

BAE Systems unveils UAVs

Herti, Raven and Corax

Having worked on a number of unmanned projects for some years, BAE Systems revealed details of three of its fixed-wing programmes in December 2005 and February 2006. Two of the programmes are related, and embody low-observables (LO) technology, drawing on company research work that has been ongoing since at least the early 1990s and which resulted in a full-size non-flying stealth demonstrator dubbed Testbed. To underline the importance of UAV business to the company, in February 2006 it established an Autonomous Systems and Future Capability (Air) department to concentrate on UAV-related work.

Both the Corax unmanned reconnaissance air vehicle (URAV) and Raven unmanned combat air vehicle (UCAV) sub-scale demonstrators highlight BAE Systems' considerable know-how in UAV and LO technologies. They are important for the UK as the ongoing achievements provide significant leverage for future collaborative ventures. Britain had signed up as a partner to the US J-UCAS programme, but the cancellation of J-UCAS and subsequent re-evaluation of the US services' requirements highlighted the vulnerability of the UK to any changes brought about by the US. The development of robust UCAV technology will allow the UK to negotiate for significant

One of two Herti 1A prototypes shows the type's motor-glider origins. BAE Systems forecasts a sizeable market for this class of general-purpose UAV, in both civilian and military arenas. Possible civilian roles include pipeline monitoring and environmental work.

status in any future partnership – with either the US or European nations – or to go it alone if needs be.

Corax and Raven are highly unstable vehicles, requiring fly-by-wire control systems. Proving the ability to fly and control a highly unstable UAV is one of several key aims of the programmes, as previous efforts in the US (notably concerning the Lockheed Martin/Boeing RQ-3 DarkStar) have shown this area to be fraught with problems.

As well as its fixed-wing programmes, BAE Systems has also developed a small ducted-fan vertical take-off UAV that is intended to deploy the dispersed Wolf Pack unattended battlefield communications intelligence/jamming system. This system was designed and flown in the US. Back in the UK, the company is also investigating sense-and-avoid systems for unmanned vehicles, using a Jetstream as a surrogate UAV.

Herti

This is a family of low-cost, non-stealthy UAVs that can fulfil a variety of missions, both civil and military. Basic airframe design was undertaken by J&S Aero Design in Poland, which produces a

range of motor-gliders. The first version was the Herti 1D, a 350-kg (772-lb) jet-powered vehicle with an 8-m (26.2-ft) wingspan. This led to the 450-kg (992-lb) piston-powered Herti 1A, with a 41-ft (12.5-m) wingspan. The Herti is relatively conventional, with high-aspect ratio wings and a butterfly tail, spatted undercarriage and a centrally mounted pusher engine. The vehicle is capable of fully autonomous operations. Herti 1A has an endurance of more than 25 hours and an operational radius in excess of 1000 km (621 miles).

The Herti 1A has been tested from the relatively remote civil airfield of Machrihanish, near Campbeltown in Scotland. The first fully autonomous flight – the first by a UAV in UK airspace – was made on 18 August 2005. The UAV has tested a chin-mounted electro-optical sensor, and was due to trial a Selex side-looking airborne radar installation in 2006. It has also tested the BAE Systems Imagery Collection and Exploitation (ICE) camera-based system. By the end of the year BAE Systems was expecting to have completed eight Rotax-powered Hertis to complement the two BMW-powered prototypes.

Left: Wearing Class B civil registration G-8-008, a Herti 1A lands back at Machrihanish after a test sortie. BAE Systems is rapidly producing eight further aircraft so that UAV tests can be accelerated.

Below: One of the two Ravens taxis at Woomera. The aircraft exhibits classic stealthy UCAV lines, with a top-mounted, partially shrouded intake and a frontal aspect dominated by just two straight leading-edge lines.

Trials with flying four Hertis simultaneously will be undertaken, as well as rail-launched take-offs and grass strip operations.

Raven

Built at Samlesbury, the Raven is a sub-scale demonstrator for unmanned combat air vehicle (UCAV) technology. Its airframe consists of a central body combined with blended, swept wings. There are control surfaces on the wing's trailing edge and 'drag rudders' that open above and below the outer wing panels. Raven is highly unstable and uses a dual-redundant FBW system.

The first of two Raven vehicles first flew on 17 December 2003, the 100th anniversary of powered flight, away from prying eyes at the remote Woomera Range in South Australia. It was preceded by a blended-wing demonstrator called Kestrel, which first flew in 2002. Kestrel was remotely-piloted, whereas Raven has full autonomous capability to fly missions from take-off to landing without human input. The second Raven flew in November 2004.

Tests with the Ravens have fed directly into the UK Ministry of Defence's wide-ranging and ongoing studies into future airpower, initiated in the 1990s as the Future Offensive Air System (FOAS) programme. In 2005 FOAS was super-seded by the Strategic UAV Experiment (SUAVE), which is examining the roles that UCAVs might play in the deep strike/attack mission. Raven has obviously played an important part in these evalu-ations, and the successful operations with it are expected to lead to the go-ahead for a full-size UCAV demonstrator in 2006. This will also draw on the classified Nightjar programme that has explored LO technologies through ground-based demonstrators.

Corax

Taking its name from the Latin name for the raven, the Corax is a sub-scale URAV demonstra-tor that was derived from the Raven UCAV. Its unusual shape – reminiscent of the DarkStar – combines a fuselage centrebody (shared with the Raven UCAV demonstrator) and high-aspect ratio wings for high-altitude, long-endurance opera-tions in the surveillance role. The central computer and ground station is the same as that of Raven, furthering the commonality between the two vehicles. A series of four moving sections along the trailing edge of the wing provide control and extra lift, augmented by movable control

surfaces at the ends of the fuselage aft-bodies that extend either side of the jetpipe. At the time of writing no dimensions for either Corax or Raven had been released, but the wingspan of the URAV demonstrator is in the region of 10 m (33 ft).

Like the Raven, Corax undertook flight trials at Woomera in South Australia, from January 2005. It is clearly intended to demonstrate technologies for the surveillance role, and a derivative vehicle could be a candidate to (partially) fill the Project

One of the challenges of the Raven programme was to create a stealthy, stable vehicle that has no vertical fins. Unusual surfaces that open above and below the wings (as fitted to the B-2) provide differential drag to slew the aircraft one way or the other to keep the aircraft flying straight.

Dabinett requirements that are seeking a long-term replacement for the Canberra PR.Mk 9 in the strategic reconnaissance role from around 2017.

David Donald

As well as demonstrating URAV technologies, the Corax also demonstrates the modular approach that BAE Systems has adopted in producing common components (such as central fuselage, flight control system and ground equipment) for different UAVs for various roles. Below the aircraft takes off from Woomera.

Boeing 737 AEW&C

Project Wedgetail

Boeing's 737 AEW&C represents a smaller, smarter breed of AEW platform. After nearly a quarter of a century of deliberation, Australia chose the type in 1999, and it is due to enter service in 2007 to revolutionise the way the RAAF does business.

A major milestone in Australia's Project Wedgetail 737-based AEW&C programme occurred on 16 January 2006 when the first 'green' airframe arrived in Australia for reconfiguration. Boeing Australia Limited is to reconfigure the last four of the six aircraft on order at its RAAF Amberley modification facility. Two aircraft are currently undergoing systems installation and flight testing by Boeing in Seattle and the first Australian aircraft, Wedgetail 3, was immediately inducted into the programme upon arrival. The four Australian conversions are to be completed at four-month intervals, commencing in March 2007.

The acquisition of an AEW&C system has long been desired by the RAAF, beginning with a 1975 study of the USAF's new E-3A Sentry, and the US Navy's E-2C Hawkeye. Initially focused purely on air defence, analysis of requirements soon evolved to include battlespace command and control in all environments, including air defence, surveillance,

Australia became the launch customer for the 737 AEW&C, an aircraft that Boeing hopes will take a large slice of the market. To cater for the huge power requirements of the AEW&C system, large generators are fitted to the CFM56-7 engines of the 737 AEW&C. They are covered by bulges on the port side of each engine nacelle.

force co-ordination and ground support.

In 1978, Air Staff Requirement No. 64 (Air Defence Early Warning and Control Components) was issued to acquire an AEW&C capability in conjunction with other air defence elements, such as the Jindalee Operational Radar Network (JORN). By 1980 the AEW&C element had been taken out of ASR.64, becoming Project AIR 77, but little progress had been made.

Towards the end of 1985, the Australian Government revitalised AIR 77, issuing a Request For Proposal (RFP) to industry the following year. In mid-1988 the Minister of Defence announced that an AEW&C system would be acquired, signalling the release of a Request For Tender (RFT) in November. Once again. however, due to economic pressure at the time, the project was allowed to slip.

By 1994 AIR 77 had evolved into AIR 5077 Project Wedgetail and a Project Definition Study (PDS) was approved. Completed in 1996, it identified the weaknesses in the current Australian Defence Force Air Defence System (AADS), and quantified the level of performance sought. It was determined that the selected system should be capable of early warning, tracking and close control of fighter-size targets to a distance of 190 nm (352 km) in all sectors. The ESM system should have a similar capability to the ALR-2001 fitted to the RAAF's AP-3C Orions, being able to detect, track and identify air, land and maritime targets throughout 360°, as well as being fully integrated into the AEW&C mission system. In addition, the platform should have a dash speed equal to, or greater than, 280 kt (518 km/h), be able to maintain a mean line of advance of at least that speed, and operate at an altitude in excess of 20,000 ft (6096 m). Self protection would include a radar warning receiver, missile warning system, laser warning receiver and CMDS.

Following definition, the Commonwealth issued an Invitation to Register Interest (ITR) for the provision of an AEW&C capability on 28 May 1996. More than 50 responses were received, and three were short-listed. Budget approval for phase two of the project occurred on 20 August, allowing an RFP to follow on 14 November to Boeing (737-700/Northrop Grumman ESSD MESA radar), Lockheed Martin (C-130J/Upgraded AN/APS-145) and Raytheon E-Systems (A310-300/IAI Elta).

An RFT followed on 7 May 1997, and the Minister for Defence Ian McLachlan announced approval for AEW&C capability acquisition on 2 December. Initial Design Contracts, valued at A$ 8.5 million each, were awarded to Boeing, Lockheed Martin and Raytheon E-Systems on 28 January 1998, with selection of the successful contractor due in the third quarter of 1999, and Initial Operational Capability to be attained in 2003.

Winner announced

The then Minister for Defence, John Moore, announced Boeing as the preferred tenderer on 21 July 1999, with a proposal for the supply of seven aircraft, based on the 737-700IGW (the baseline aircraft for the 737-BBJ) and fitted with the Northrop Grumman Multi-role, Electronically Scanned Array (MESA) radar. Final contract signing was anticipated in early 2000 for initial delivery a little later than previously planned, in the 2004/5 timeframe. The 737/MESA combination was considered the clear winner at the end of the day, though it was acknowledged that any of the contenders would have fulfilled the requirements.

Above: The first 'green' airframe for Wedgetail conversion is rolled out at Boeing's Renton plant on 31 October 2002. The aircraft was in standard Boeing Business Jet (737-700IGW) configuration at this point.

Below: Conversion of the first two Wedgetail aircraft was undertaken at Boeing Field in Seattle, where the Northrop Grumman MESA radar was mated to the strengthened upper fuselage.

Partnering Boeing are Northrop Grumman's Electronic Sensors and Systems Division (ESSD), Boeing Australia, BAE Systems Australia, and Hawker De Havilland (a Boeing Australia subsidiary). Boeing Australia has overseen system engineering and aircraft modification support, and also has the lead in the product support and ground support systems teams. BAE Systems

Above: Wedgetail 1 takes off from Boeing Field on an early test flight. The CFM56-7 engines give the 737 AEW&C a spritely take-off.

Below: The addition of the MESA and ventral strakes required a full aerodynamic evaluation campaign. Here Wedgetail 1 undergoes flap testing.

Australia provides the Passive Surveillance System, Electronic Warfare Self Protection (EWSP) system, Operational Mission and Mission Support Segment, and the AEW&C Support Facility.

A contract for four aircraft and three options was signed on 20 December 2000, and by then the programme had matured enough to quote an in-service date of 2007 for the first two aircraft. Total value of the project was estimated at A$ 3 billion, though further budgetary restraints soon saw the third option quietly scrapped, and the project moved forward on the basis of four firm orders for aircraft, but six sets of AEW&C systems.

Preliminary Design Review was completed in July 2002, one year after the commencement of the Systems Acquisition contract, clearing the way for the release of engineering drawings and allowing the fabrication of parts to begin. Roll-out of the first 'green' airframe at Boeing Renton on 31 October 2002, before the new Defence Minister Robert Hill, marked a significant milestone. Though systems installation and integration were still some way off, several major hurdles had recently been cleared and the project was then some six months ahead of schedule.

Following on from the roll-out, Northrop Grumman's Electronics Systems Sector announced that it had completed the first MESA radar and IFF unit, and was undertaking integration and pattern testing on the company's test range. Testing was complete by mid-year and it was installed on the aircraft on 23 November, with a first flight planned for May 2004.

Options taken up

On 12 May 2004, following the Critical Design Review, Australia exercised its options to purchase the two additional airframes in a deal worth up to A$180 million. The new orders allowed greater Australian Industry Participation (AIP), to carry out conversion of aircraft nos 3 to 6 in Australia.

Australia has a 40-strong resident project team on-site at Boeing, initially headed by Air Vice Marshal Norm Gray, and Australians are also part of the Boeing Flight Test Team, gaining experience on the aircraft prior to the training of the first RAAF crews. No. 2 Squadron was chosen to operate the Wedgetail, being reformed for the purpose in January 2000. The unit had last flown the Canberra, and battle honours include opera-

The Systems Analysis Laboratory at Brisbane, which simulates air command and control operations, has been updated to include MESA data from Wedgetail, integrating it with other information from ground radars and facilities.

tional sorties from Phan Rang during the Vietnam War. A purpose-built squadron headquarters at Williamtown opened on 5 March 2004.

An AEW&C Support Centre (ASC) has been built at Williamtown, adjacent to the No. 2 Squadron facilities, and handed over two months ahead of schedule, in December 2004. BAE Systems commenced Operational Missions System (rear crew trainer) installation at the facility in mid-2005. BAE Systems are contracted to provide two fixed, and two deployable Mission Support Segment (MSS) units, and an Operational Mission Simulator. Boeing is Prime Contractor for supply of the flight simulator, and has subcontracted to Thales.

Wedgetail described

The Australian programme is named after the indigenous Wedgetail eagle species; one of the largest in the world, known for its sharp eyesight, ability to loiter at high altitudes and aggressive defence of its territory. Wedgetails are built as standard Boeing 737-IGW (BBJ) airframes on the Renton line, before flying to Maryland for fitment of auxiliary fuel tanks by PATS Inc. Wedgetail aircraft have the Baseline configuration of five auxiliary fuel cells: four in the aft hold, and one forward. Up to seven cells can be fitted if required by future customers.

Structural conversion includes the removal of the Section 46 upper lobe, necessitating the removal of over 6,000 fasteners, and replacement with a strengthened structure to support the MESA support pedestal and antenna. Numerous communications antennas are also installed at this point, including additional HF antennas in each ventral fin.

An air-to-air refuelling receptacle features in the Baseline design, installed aft of the flight deck. A 'bolt-on' probe system had been considered by the RAAF, but rejected because of the low transfer rates and increased flight deck noise associated with the system. To cope with the huge demand for electrical power, each engine is fitted with a 190-kVA Integrated Drive Generator (IDG). The increased diameter of this unit means that each left-hand fan cowl incorporates a blister.

Airborne Mission System design is based on open-system architecture, utilising a high proportion of Commercial Off The Shelf (COTS) components. Information is displayed on 10 iden-

Above: The Rocky Mountains of Washington state form an impressive backdrop as the first Wedgetail cruises on an early flight. The aircraft trails a drogue from the top of its fin for air data measurement.

While conversion work has shifted to Australia, trials are continuing from Boeing Field. Here the first two Wedgetails are seen at the Boeing plant, during the brief period when both were flying before no. 1 (below) entered the shop for fitment of its mission systems.

Wedgetail walk-round

Right: Perched atop the 737's spine is the low-drag 'Top Hat' structure housing the MESA sensor group, which includes 288 transmit/receive modules. The radar has side-facing modules in the vertical structure, and others in the upper fairing. The IFF system shares the same aperture as the radar. Unlike earlier fixed-array AEW radars, the MESA can transmit signals fore and aft, meaning that it can provide 360° coverage, thereby maintaining a radar picture even when the aircraft is turning through 180° at either end of a racetrack orbit.

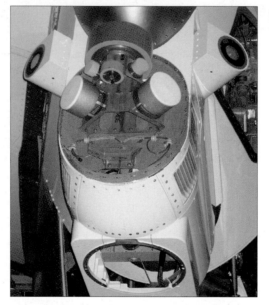

Left: The chin radome houses part of the ALR-2001 suite. Behind it are outward-facing sensors for the AAR-54 missile warning receiver system.

Above and left: The tail sensor group is mounted above and below the aircraft's auxiliary power unit exhaust. It comprises JTIDS antenna (grey radome at top), ALR-2001 (black radome) and AAR-54 sensors facing obliquely outwards. The orange fixture is a tail bumper fitted for flight tests, but indicates the position of the AAQ-24 Nemesis DIRCM.

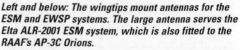

Left and below: The wingtips mount antennas for the ESM and EWSP systems. The large antenna serves the Elta ALR-2001 ESM system, which is also fitted to the RAAF's AP-3C Orions.

The Wedgetail's comprehensive communications suite is served by forests of antennas above and below the fuselage. Many are grouped in a pallet-mounted 'farm' (above). Satcoms systems are served by the T-shaped aerials on the spine (below). TACAN and GPS antennas are flush-mounted at the front.

---

tical multi-purpose consoles in the main (forward) cabin, each capable of displaying all information and performing every task. Australian aircraft will carry a normal complement of seven mission specialists, and a six-person crew rest facility and galley is installed in the mid cabin.

Communications equipment includes SATCOM, three HF radios (one antenna in each ventral fin and one in the fin leading edge), VHF/UHF Comms and a Rockwell Collins Joint Tactical Information Distribution System (JTIDS) Link 16 datalink.

The heart of the system is the MESA radar with integrated IFF and practically the entire aft cabin is given over to radar support electronics, including power supply, manifolds, transmit/receive modules, and processor cabinets. Brochure specifications include an unclassified detection range of 400 km (216 nm) against an aerial target, and 300 km (162 nm) against a surface target. Radar scan times are selectable between three and 40 seconds. IFF scan times are selectable between 10 and 40 seconds.

The fuel jettison system allows the Wedgetail to get down to landing weight and return to base quickly in the event of a major emergency such as an engine failure.

Electronic Support Measures (ESM) are provided by an Elta ALR-2001 installation, with antennas in each wingtip, in a fairing under the nose radome, and in a specially-designed mounting beneath the APU exhaust. The EWSP systems consists of a Northrop Grumman AN/AAQ-24(V) Nemesis Directional Infra-Red Countermeasures (DIRCM) turret in a fairing beneath the tailcone, manufactured in Australia by Hawker De Havilland, Elisra LWS-20(V) Laser Warning Sensors and Northrop Grumman AN/AAR-54(V) Missile Warning System Units. BAE Systems AN/ALE-47 Countermeasures Dispenser Systems (CMDS) are installed on either side of the lower forward fuselage.

Test programme

Wedgetail 1 made its first flight from Boeing Field on 20 May 2004, eight days ahead of schedule, with Boeing pilot Charles Gerbhart comment-

ing, "The plane handled beautifully, just like a 737-700. It was very similar to our experience in the flight simulator. There weren't any surprises". This aircraft was the aerodynamic test vehicle, designed to validate the handling qualities of the modified aircraft, before entering the hangar at Boeing Field in late 2005 for Mission Systems equipment installation.

Testing progressed relatively smoothly, and included close-proximity trials with tankers – up to contact distance – using both the KC-135R and KC-10A (fuel transfer testing will be undertaken in 2006). Both these, and hot weather and Rejected Take-Off (RTO) trials, were carried out at Edwards AFB. Initial flight-testing progressed ahead of schedule, Boeing claiming that the aircraft enjoyed double the per-day sortie rate previously achieved by the company.

Right and below: The Australian taxpayers got their first chance to see the Wedgetail up close in March 2005 when the first aircraft undertook a tour of Australian bases and participated in the major biennial airshow at Avalon, near Melbourne.

Though not fitted with mission equipment, power distribution systems for the MESA radar were in place, allowing operation in test modes via lap-top computer. During the trials, early high-

Above: The 'Top Hat' radar is mounted slightly nose down so that it 'flies' with the lowest drag at the Wedgetail's normal operating speed. Note that most of the cabin windows have been blanked off.

Boeing's System Integration Laboratory in Seattle, with Build No. 5 successfully completed in October 2004. Software Build No. 6 testing began the following month, and Boeing's Vice President of AEW&C Programmes, Patrick Gill, says that software testing is "generally" on schedule, with over one million lines of code under test by late 2004. Software Build Nos 7 and 8 for Wedgetail 2 followed in mid-2005.

Wedgetail 2 is the first to fly with the full mission electronics suite, and a small 'window' allowed both aircraft to operate together for a short period of time, prior to Wedgetail 1 being laid-up for Mission Systems installation. Wedgetail No. 2 is also the first to be fitted with an operable fuel-jettison system (the first aircraft initially flew with only an aerodynamic representation of the discharge tube fitted), and is the first 737 variant to be offered with this feature. Although the design does not require fuel jettison, the RAAF requested a system to allow greater safety margin in case of engine failure on take-off from forward operating locations, such as Tindal or Darwin.

In November 2005 Wedgetail 2 was flown to Seoul to participate in the Korean Air Show. During the Pacific crossing, Boeing and RAAF pilots used the opportunity to test the mission

Wedgetail 2 was the first to be fitted with the mission systems, conducting a first MESA radar trials sortie on 1 September 2005.

power radar and EMI testing was carried out by Northrop Grumman at its Victorville, California, testing range, commencing in late 2004. These trials highlighted a shortfall in expected radar performance, resulting in changes to the antenna cross-section – "a thicker surfboard on the roof", as one Boeing test pilot described it.

Flight test progress allowed Wedgetail 1 to visit Australia during March 2005, visiting several RAAF bases and industry facilities around the country. During this time it stole the show at the Australian International Air Show, held at Avalon.

Mission System software testing continues at

Wedgetail 3 arrived at RAAF Amberley in fine style on 16 January 2006 (right). While the 737 AEW&C represents the RAAF's future, the long-serving Caribou in the background (below) is a reminder of the service's proud heritage.

computing software's ability to control the HF radios, thereby minimising disruption to the test schedule.

Radar and Mission Systems Integration testing is continuing, with the first airborne testing of the MESA radar occurring during a six-hour flight over Washington State on 1 September 2005. Declared, "a complete success" by Boeing and Northrop Grumman engineers, it followed a three-week period of ground testing which verified compatibility with other aircraft systems while operating and scanning through 360°.

Australian integration

The decision to purchase two additional aircraft in May 2005 has meant that Wedgetails Nos 3 to 6 will now be completed by Boeing Australia, utilising about 170 local personnel. Five technicians from Seattle are in Australia to assist in the conversion of Wedgetail 3, and Boeing Australia personnel have participated in the conversion of earlier aircraft in the USA. According to Patrick Gill, all six aircraft will be either in modification or test by the end of 2006.

Wedgetail No. 3 delivery to the RAAF is currently planned for March 2007, with the remainder following in August, December, and March 2008 respectively. No. 2 Squadron will accept the first two aircraft in January 2007, slightly later than the December 2006 contract date due to the recent machinists' strike at Boeing. Despite this, Initial Operational Capability (IOC) will be achieved in July 2007, by which time the RAAF will have four aircraft and four trained crews on line. IOC is based upon the ability to have an aircraft on station, 24 hours a day in an operational environment.

Left: The Wedgetail will be a cornerstone of Australia's air defences for several decades and a key node in the move to netcentric warfare. As well as traditional AEW tasks, it will also undertake expanded command post/battle control duties.

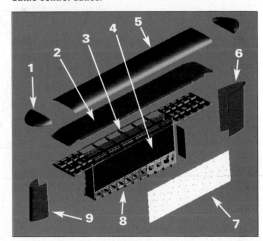

Right: This Northrop Grumman graphic shows the major components of the MESA/'Top Hat' installation.

Key
1: Forward bull nose
2: Top Hat manifold
3: Top Hat structure
4: Columns
5: Top Hat radome
6: Aft fairing
7: Side array radome
8: Base beam
9: Forward fairing

737 AEW&C specification

Maximum taxi weight: 171,500 lb (77792 kg)
Maximum take-off weight: 171,000 lb (77566 kg)
Maximum landing weight: 134,000 lb (60782 kg)
Operating altitude: 41,000 ft (12497 m)
Nominal operating altitude: 30,000 to 40,000 ft (9144 to 12192 m)
Nominal cruise speed: 360 kt (667 km/h) TAS
Dash speed: 473 kt (876 km/h) TAS
Mission endurance: 10 hours (unrefuelled), 16 hours with AAR
Endurance range: 3,500 nm (6481 km), unrefuelled

Performance figures are brochure values, courtesy Boeing

performing all normal tasks, though a specialist ESM operator will be included in each crew for passive analysis. To meet forward-deployment needs a purpose-built hangar, capable of housing two aircraft, will be constructed at Tindal.

Boeing Australia has recently completed an A$3 million upgrade to its Brisbane Systems Analysis Laboratory (SAL) to provide Wedgetail modelling. Named Audax (the Wedgetail genus name), the simulation has been constructed from MESA radar test bed data. The technology allows integration of data from HF, VHF, UHF, SATCOM and Datalink communications networks, from JORN, and Ground-based Air Defence Radar simulators, and various microwave Air Traffic Control radars.

Peace Eagle for Turkey

Further success for Boeing came in June 2004 when Turkey placed a US$1 billion order for four aircraft, under the Barifl Kartali Programi (Peace Eagle programme). The first airframe rolled off

Eight Australian Mission Systems personnel have been flying aboard USAF E-3Bs with the 552nd Air Control Wing at Tinker AFB Oklahoma for the past two years, and a further four with the RAF Sentry force at Waddington. In addition, 10 people are embedded in Boeing as part of the Wedgetail flight test crew. All will be back in Australia by years' end, and the time between delivery of the first two aircraft and IOC will be occupied by the work-up of maintenance personnel and flight crews. The combination of Wedgetail system experience and AWACS tactical knowledge will provide the basis for war-fighting capability to be achieved at IOC.

Australia's six-aircraft AEW&C fleet will be based at Williamtown alongside the bulk of the RAAF's fighter force. Regular detachments will be undertaken to the northern airfields such as Darwin and Tindal, as this is the principal area of operations

Mission crew will be a mixture of RAAF and RAN personnel, with (Ground) Fighter Controller Qualifications as the core skill. Each crew member will be capable of manning all stations and

Boeing 737-7ES AEW&C production

No.	MSN	Line no.	US reg.	Customer	Serial	Conversion
1	33474	1245	N378BC	RAAF	A30-001	Boeing Field
2	33542	1232	N358BJ	RAAF	A30-002	Boeing Field
3	33962	1614	N356BJ	THK		Boeing Field
4	33476	1810	N359BJ	RAAF	A30-003	Amberley
5	33963	1839	N360BJ	THK		Ankara
6	33964*		N362BJ	THK		Ankara
7	33965*		N367BJ	THK		Ankara
8	33477*		N361BJ	RAAF	A30-004	Amberley
9	33986*		N363BJ	RAAF	A30-005	Amberley
10	33987*		N364BJ	RAAF	A30-006	Amberley

* Exact order of production unknown at the time of writing

the Renton assembly line on 9 November 2004 and is currently undergoing conversion in Seattle.

Turkish aircraft will differ from baseline (and Australian) aircraft in that they will have a Turkey-specific ESM suite, produced by Elta, and radios that are specific to local requirements. The remaining three aircraft on order will be completed by Turkish Aerospace Industries (TAI) in Ankara. Delivery of the first aircraft is due in July 2007, with the last to be handed over to the Turkish Air Force in June 2008.

Future sales

South Korea has been engaged in its own E-X AEW&C programme in recent years, and Boeing's 737 AEW&C is competing with an IAI/ Gulfstream Aerospace team, based around the Gulfstream 550 corporate jet design, fitted with Elta's Phalcon radar.

The Korean RFP, for four aircraft, was released in March 2004 and a Korean Test and Evaluation Team visited Seattle in September. During this time, three ROKAF pilots had the opportunity to fly the Australian Wedgetail, and Boeing held pre-contract discussions with the Korean Government. Source selection was to occur in December 2004 with Boeing the front-runner, after the Gulfstream/Phalcon bid did not meet specification. However, the Korean Government considered that a sole-source contract contravened procurement policy and the contest is being re-run. Current thinking suggests that down-

While Wedgetail 1 receives its mission system in the background, the first 737 AEW&C for Turkey undergoes conversion in the Boeing Field facility. Here the MESA array is fitted at the beginning of March 2006. On 13 March Boeing delivered the first green airframe to TAI for conversion in Turkey.

selection will now occur in May 2006. The original Boeing proposal would have seen delivery between October 2009, and April 2011.

Though Korea is presently the main export focus, a campaign is underway to supply both AEW&C and P-8A Multi-Mission Maritime aircraft to Italy (possibly under a single contract, due to Italian defence spending restrictions), and Boeing considers that potential exists for sales to the United Arab Emirates and possibly Spain.

In the Asia-Pacific region, Boeing sees an "enduring need" for a replacement AEW&C type in Singapore, and for capability in Malaysia. Market projection is for about 30 aircraft across the globe, although the 737 faces competition from the Saab 2000 and EMB-145, both with Erieye radar, and the E-2 Hawkeye.

Nigel Pittaway

Ceibo 2005

In November 2005 the Fuerza Aérea Argentina hosted a major South American exercise – the first of its kind to be undertaken without US or European participation.

From 12 to 27 November 2005 Argentina hosted the international exercise Ceibo 2005, centred around El Plumerillo air base at Mendoza, with a second base at Villa Reynolds, some 400 km (250 miles) to the southeast. Mendoza airbase is situated in a well-known wine-producing area in the mid-west of Argentina, along the foothills of the Andes mountains. Besides a large contingent of Argentine aircraft, fighters from the air forces of Brazil, Chile and Uruguay also took part.

Argentine Air Force Colonel Miguel Antonio Cruzado, Chief of Staff of DIREX and Commander of the Air Operations, explains about Ceibo 2005, "The name Ceibo is taken after our national Argentine rosé red flower. The operation follows the NATO model in which Coalition Forces participate in a UN Peace Mission. Concerning air operations in general, the exercise comprises two weeks in which the first week is used for familiarisation between the participant air forces, and is then stepped up towards war with actual COMAO missions. After the weekend the exercise is in the war status with two COMAO waves flown per day.

"Of course, much more happens in the background, like the planning and execution of the missions by the Joint Force Air Component Command (JFACC). JFACC receives its orders from the Direction and Controlling Staff, which runs the war game and also supplies virtual infor-

mation for the media. For the latter a delegation of communications students from Cuyo University was working at PC stations at Mendoza, acting as if they were the media of the virtual countries. 'Canopus', south of Mendoza in San Rafael province, is the aggressor. 'Canopus' has invaded 'Orion', situated to the north of Mendoza in the San Juan province. A third country, 'Alpha Centauro', the Mendoza province, was the host for the coalition forces. Interestingly, the exercise sheets showed Russian-built aircraft as the 'Canopus' fighter force."

Proceeding the action of Ceibo 2005 was a paper exercise concerning United Nations resolu-

As host nation Argentina provided the bulk of the aircraft for Exercise Ceibo. At the heart of the ground attack forces were seven A-4AR Fightinghawks (above), plus a single TA-4AR two-seater, from Grupo 5. The 'Red Air' opposition was provided by the Mirage IIIEAs of II Escuadrón, Grupo de Caza 6 (below). The Mirages carried a single MBDA Magic 2 acquisition round and operated from Villa Reynolds.

tions and approval for intervention. In the end, Resolution 013/05 was agreed upon by the UN Security Council, and Task Force 013/05 was formed operating from Mendoza in the virtual country of 'Alpha Centauro'.

"We feel it is very important for South American Air Forces to work together," says Colonel Cruzado, "and we are particulary proud of the fact that Ceibo 2005 was the first international Latino exercise that was organised and prosecuted by our own means, without support from outside South America. Due to our commitment in previous Aguila, Cruzex and Salitra exercises, we have managed to improve our support significantly – especially the Argentine-developed computer network has proved to be an advantage. We used to have a planning time for Air Task Orders (ATO) of 72 hours, but we now manage to get an ATO out in 48 hours."

Time was also gained by the preparation of the exercise itself, as Ceibo was put together in a relatively short timeframe. It was Argentine Air Force Brigade General Roles who initiated the idea for Ceibo during the last Salitra exercise in Chile

Resplendent in its new low-visibility grey scheme is an IA-58A Pucará from I Escuadrón, Grupo 3 de Ataque.

October 2004. The first meeting of partipants of Ceibo 2005 started in June 2005, followed by a second meeting in September 2005. Beside the main participants, observers from Paraguay and Venezuela were also present.

On the question of how the countries worked together on Ceibo 2005, Colonel Cruzado smiles, "Very, very good! For instance, in the first week Argentine Mirage 5s flew air-to-air missions together with Chilean Mirages 5s without any problems. For the language we used English or Spanish. Our Portuguese-speaking Brazilian participants get along with that fine."

In exercises like Ceibo the asset of AWACS is usually appreciated for the overview and assessment of air operations. Initially it was planned that a Brazilian R-99A AWACS would take part in

The Mirage still rules in the southern part of the American continent. This pair comprises a I Escuadrón Mirage 5PA Mara and a II Escuadrón Mirage IIIDA, both from Grupo 6 at Tandil.

Ceibo but this support was withdrawn. Possible reasons were rumoured to be the 24-hour per day surveillance required during visit of President Bush to Mardel Plata, while one of the Brazilian R-99As was away on a sales/demonstration tour to the air shows in Dubai and Malaysia. About the missing AWACS, Colonel Cruzado remarks, "We get our radar picture from four combined radar sources. We have the civilian ATC radar of Mendoza and three mobile Westinghouse TPS-43 radars based at San Juan, San Rafael and San Luis which gave us radar coverage of, respectively, north, south and east of Mendoza. The west of the airspace is naturally bordered by the Andes. This works out well."

First week

During the first days the participants meet each other and make familiarisation flights in the back

Grupo 6's Mirage 5PA Maras operate alongside the Fingers in I Escuadrón. They have RWR antennas either side of the nose, and chaff/flare dispensers scabbed on under the jetpipe.

A combination of APG-66 radar and AIM-9M Sidewinders make the A-4AR useful light fighters, in addition to their attack role.

seats of each other's aircraft. Mirage 5 pilot of Grupo 6, Lieutenant Herberth Rinaldi, commented after a familiarisation flight in a Pucará, "I thought we flew low with our Mirages but with the Pucará it is amazing how fast the world goes by at two metres height! My instincts wanted it to just get up, up, up. Really an exciting trip!"

This phase is followed by further Forces Integration Training (FIT) and Flight Affiliation Missions (FAM) to ensure the safe execution of the exercise. Also in the first week, Nationals Evacuation Missions (NEO) are carried out by dropping Argentine commando forces to evacuate an 'Alpha Centauro' minority living in the invaded 'Orion' and threatened by 'Canopus' forces. The dropping and the evacuation was performed by a Fokker F27 Friendship transport of Grupo 9/Esc. 6, normally based at Comodoro Rivadavia

Grupo 6 operates the FAA's fleet of IAI Finger As. This Israeli-built Mirage version can be distinguished by having an angular nose with strakes either side.

in the far south of Argentina. The commandos were also tasked with real-life guarding of the ramp that accommodated the coalition fighters.

Towards the end of the first week the exercise proceeded towards war with actual COMAO missions. At the weekend a public airshow was organised with a static display of all types of fighter aircraft, and several display flights. The Mendoza-based Escuadrilla Acrobatica flew routines with the Su-29AR aerobatic display aircraft.

LIVEX

During the first week the pressure slowly built up, but in the second week the war was fully on. For this purpose the aggressor forces, comprising Mirage IIIEAs and one Mirage IIIDA, were relocated from Mendoza to Villa Reynolds in 'Canopus'. Monday started with air supremacy,

which was dedicated to win air superiority and diminish the enemy's air capability. From Tuesday onwards only air-to-ground missions were flown by 'Alpha Centauro'. Fighter Bomber Attack (FBA) was performed by the Argentine Mirage 5 Fingers/Maras and A-4AR Fighting Hawks, as well as Brazilian A-1A/Bs. Close Air Support (CAS) was taken care of by the 'slow movers' group comprising Argentine IA-58A Pucarás, IA-63 Pampas and MS-760 Moranes, and Uruguayan A-37B Dragonflies. The interdiction strike task (IDS) to reduce the enemy's re-supply capability was performed by the Chilean Mirage 5MA/MD Elkans. Each day 40 to 45 missions were planned, of which 95 per cent were actually flown.

Argentina

The host supplied the largest contingent of personnel and aircraft. In total, 750 personnel were needed to operate 46 aircraft. The main supplier was Grupo 6 de Caza, normally based at Tandil, but operating from Mendoza with 13 aircraft: four IAI Fingers, two Mirage 5A Maras and seven Mirage IIIs. The second squadron of Grupo 6 performed the task of aggressor in the air-to air role with six Mirage IIIEAs and one Mirage IIIDA operating from Villa Reynolds air base. Their task was to intercept the FBA and IDS missions from 'Alpha Centauro'.

Eight A-4AR Fighting Hawks, including a TA-4AR two-seater, were delegated from Grupo 5 de Caza at Villa Reynolds. These A-4ARs proved to have a high level of serviceability and brought a feeling of the past to Mendoza, as this airfield was once a major Skyhawk base.

Close Air Support was performed by six IA-58 Pucarás of Grupo 3 de Ataque based at Reconquista. Beside the bare metal finish some

Pucarás wore the new overall light grey colour scheme. This new camouflage was added in combination with an instrument update. The Pucará fleet is currently awaiting approval for an engine upgrade. The CAS task at the home base has now been converted to interception of smugglers and drug traffickers from Brazil and Paraguay. So far only interceptions have been made, but the Grupo is awaiting the outcome of a political proposal for permission to shoot the invaders down.

Mendoza is the home base for Grupo 4 de Caza, tasked with combat training. Escuadrón 2 is equipped with the Morane-Saulnier MS-760 Paris. This license-built version of the Paris can be armed with two 7.62-mm machine-guns in the nose and can carry bombs under the wing pylons. The old Morane is about 47 years old and the fitting of a Garrett GPS-3 system is in sharp contrast with the analog instruments.

During the familiarisation phase serious dogfights with Pampas were staged, performing

four set-ups in which up to 4 g was pulled, the limitation for the Morane due to the wing tank configuration. In LIVEX pairs of Moranes flew CAS missions, dive-bombing from 12,000 to 600 ft (3658 to 183 m).

In Argentina advanced combat training is performed with the IA-63 Pampa. Currently the Pampas are undergoing an update programme at FMA in Cordobá. This comprises the fitting of two TFT displays and HOTAS, among other updates. First the Pampas stored at FMA since 1990 will undergo the update, followed by another ten, all leaving FMA as AT-63s. Due to this programme only three Pampas were present at Mendoza (and four surviving Moranes). Three Moranes and three Pampas took part in the 'slow movers' part of Ceibo 2005.

For transport duty a single C-130B of Grupo Aereo 1 de Transporte/Escuadron 1, from Buenos Aires, was stationed at Mendoza wearing a new green/brown tactical colour scheme. Operating on a daily basis in Ceibo 2005 was

The Brazilian contingent operated five single-seat A-1As (above) and one two-seat A-1B (left). The AMXs carried inert examples of the little-seen MAA-1 Piranha missile. In February 2006 Brazil announced the Piranha would be replaced on its A-1s and upgraded F-5s by the South African Denel A-Darter.

TC-69, a KC-130H supplying fuel for Argentine A-4ARs and Uruguayan A-37Bs. For paradropping and transport one Fokker F27 Friendship transport of Grupo 9/Esc. 6 deployed to Mendoza from Comodoro Rivadavia. SAR duty was performed by two locally based SA-315B Lamas, especially fitted for rescue in the high Andes, and one Bell 212.

Chile

The Chilean contingent brought 76 personnel to Mendoza to support the Elkans of Grupo 8, normally based at Antofagasta. Elkan means 'Guardian' in the local Mapute Indian language.

The aircraft are former Belgian Air Force Mirage 5s modified to MirSIP standard and delivered to Chile in 1995. During the first week the large 1700-litre (449-US gal) wing tanks of the Elkan were switched for the smaller 500-litre (132-US gal) wing tanks for an air-to-air mission every morning. Normally, the Elkan is tasked only with interdiction missions for which extended range is needed, hence the large wing tanks.

"We are glad to attend Ceibo 2005," says Squadron Commander of Grupo 8, Lieutenant-Colonel Ibacache Claudio Salinas, "It is our first time to be in Argentina with our Elkans. A long time ago the situation was not so friendly, but now

we are happy to feel very welcome." Concerning the 1 vs 1 and 1 vs 2 fights against the Argentine Mirages and A-4ARs, Colonel Salinas smiles and states "Well, it seems we are more or less alike. Pilots are pilots. Both nations fly the Mirage very professionally and by the book without any tricks. Sometimes the Chileans make a good move and sometimes the Argentinians. In the end we have about the same level of experience. Due to the altitude [761 m/2,497 ft above sea level] and the high temperatures [over 30°C/86°F] the flying is different for us and we have to be careful." When fitted with the large tanks the Elkans had little runway length to spare during take-off.

Brazil

Brazil dispatched 78 personnel, five EMBRAER A-1As and one two-seat A-1B to Ceibo 2005, all from Grupo 10 at Santa Maria. All but one of the AMXs were painted in the new standard green lizard camouflage. The AMXs wore the squadron badge – a red lion – on the tail,

The ramp bustled with activity during Ceibo as the participants launched COMAO attack missions. Several years ago the idea of Argentine and Chilean warplanes operating side-by-side would have been unthinkable

and all aircraft were fitted with two inert examples of the indigenous Mectron/CTA MAA-1 Piranha air-to-air missile. The A-1s are all fitted with air refuelling probes and joined up with the KC-130H for hook-ups during missions, although they only refuelled when it was needed.

Uruguay

For the Uruguayan A-37Bs and 30 personnel of Grupo 2 at Durazno it was the first chance to participate in an international exercise. For Ceibo 2005 three A-37Bs of Grupo 2 were planned to do air refuelling with the Argentine KC-130H Hercules tanker. As the Uruguayan pilots had no

experience with air refuelling, a special pre-exercise – Tanque 2005 – was held on 28/29 September 2005, in which an Argentine KC-130H flew to Durazno to perform ten air refuelling training missions.

Squadron Commander Lieutenant-Colonel Jose Visconti recounted that, "On 18 November we had the first hook-up at Ceibo 2005. This was done in a ceremonial way by an A-37B flown by me, accompanied by the youngest pilot of the squadron. In every COMAO mission we flew a two-ship that each performed air refuelling. During Ceibo we flew a total of 80 hours. Our A-37Bs are not high performance, but we do form

Grupo 4 de Caza – the unit that performs the advanced training role at Mendoza – also participated in the exercise in the light attack role with three Pampas and three MS-760s.

part of the package. Being part of Ceibo and meeting people gave us lots of opportunities, and are starting points for future activities."

Cees-Jan van der Ende and René van Woezik

Ceibo represented a milestone for the Fuerza Aérea Uruguaya as it was the FAU's first multi-national exercise. The FAU's A-37B Dragonfly fleet is in the process of being overhauled, with work performed by ENAER in Chile.

Last cruise of the Tomcat

After three decades the US Navy service of the mighty Tomcat is drawing to a close. The F-14 ended its front-line duty as it began – on combat operations. For the last cruise of the 'Cat' two squadrons deployed aboard USS *Theodore Roosevelt* for operations over Iraq.

Below: VF-213's most colourful aircraft wears the appropriate Modex '213'. The squadron also has a 'CAG-bird' and 'Boss-bird' with full-colour markings, but not the blue rudders. Most of the Tomcats of VF-213 and -31 are around 14-16 years old. The first production F-14D flew in February 1990, while the last of 37 was delivered in July 1992. Grumman converted another 18 F-14As to D standard, of which only a few are still flying.

Seldom in the history of aviation has the withdrawal of an aircraft evoked so much emotion as the rapidly approaching retirement of the Grumman F-14 Tomcat from US Navy service. In September 2005 the very last Tomcat cruise commenced when VF-31 'Tomcatters' and VF-213 'Black Lions' set sail on the USS *Theodore Roosevelt* (CVN 71) for a scheduled six-month cruise towards the northern Arabian Gulf. The latter transitioned to the Boeing F/A-18F Super Hornet in spring 2006, with the former converting to the F/A-18E Super Hornet in the summer.

Many feel the F-14 Tomcat has been retired too soon, and not without reason. This magnificent jet is still an outstanding aerial combat performer and outclasses the newer Boeing F/A-18C Hornet in many aspects. Even over the F/A-18E/F Super Hornet – the Tomcat's successor – the swing-wing jet has certain advantages that the US Navy will unquestionably miss in future combat.

There has been a lot of discussion about this issue. Clearly, the Tomcat is a lot faster, has a better climbing rate, a higher payload, a better bring-back capability and endurance. It is a perfect strike and close air support (CAS) platform. The F-14 was the only fighter capable of employing the long-range Hughes AIM-54 Phoenix air-to-air missile, a missile hardly fired in combat by the US Navy. Nevertheless, the Iranian Air Force reportedly used this missile with excellent results in the Iran-Iraq war during the 1980s. The US Navy retired it in 2004 and the Tomcat now solely relies on the short-range Raytheon AIM-9 Sidewinder and medium-range Raytheon AIM-7 Sparrow.

Although exclusively used as an air-to-air and reconnaissance platform for the first two decades of its career, the Tomcat has starred as a fighter-bomber since the mid-1990s. Its excellent range and weapon load made it a great jet to cover some of the capabilities that were lost with the retirement of the Grumman A-6 Intruder attack plane.

Initially the Tomcat carried dumb bombs but soon some intrepid naval aviators recognised its potential as a precision bomber. A Lockheed Martin LANTIRN targeting pod was added and laser-guided bombs soon became standard ordnance. Its two-man crew also made it ideal for duties that had become unknown in the US Navy: the Forward Air Control-Airborne, or FAC-A mission. This specialised task is demanding. The identification and targeting of small targets close to friendly forces is difficult and the crew has to communicate extensively with ground troops, as well as with other strike aircraft. A qualified two-man crew is therefore highly desirable. The only two Tomcat squadrons flying in Operation Allied Force over the former Yugoslavia in 1999 became famous because of their achievements in the close air support role.

Soon after the JDAM satellite-guided bombs became available the F-14B and F-14D were certi-

VF-31's black-tailed 'CAG' jet flies over Iraq in November 2005. It carries a single Sparrow on the port wing to provide a measure of self-defence, and to partially balance the LANTIRN pod on the starboard side.

fied to employ this class of weapon (the older F-14A did not have the required digital databus). JDAM received its baptism of fire with the Tomcat in 2002 during Operation Enduring Freedom in Afghanistan, where the long range, high payload and two-man crew made it the Navy's weapon platform of choice while operating from carriers in the Indian Ocean.

However, potential combat performance is not the all-deciding factor when it comes to keeping or replacing an aircraft type. The costs of modern warplanes and aerial warfare have exploded during

Armed and ready, a VF-31 F-14D waits in front of the jet blast deflector to move forward to the catapult. The F-14D was the only Tomcat model to sport both infra-red and AXX-1 TCS under the nose.

the last decades. Modern air forces have to look at service life and maintenance costs in order to determine the feasibility of purchasing or keeping aircraft. And the Tomcat is a bad performer when it comes to maintenance. "It is one of the greatest planes ever built for the Navy, and also one of the most beautiful," contends Captain William G. Sizemore II, commander, Carrier Air Wing 8 on the *Roosevelt*. "It has great endurance and flexibility. It is bittersweet to see it go but it is getting old, it is three times as expensive to fly as the Hornet."

Maintenance

Commander James 'Puck' Howe, as XO the second-in-command of VF-31, says the mission-capable percentage rates of his Tomcats are in the high 80s. "That rate tells that we can fly our mission but not necessarily that every system is working properly. However, the full mission-capable rate, which means that everything works, is in the low 80s which is still very good." These are fairly impressive numbers but it does not reflect how much effort is required in order to

Hook down, VF-213's 'CAG-bird' enters the pattern for **Roosevelt***. The squadron modified its lion markings to have two tails when the squadron transitioned to the twin-tailed Tomcat in 1976.*

achieve them. In VF-31 Lieutenant John Turner, the Maintenance Material Control Officer, is responsible for the maintenance department. "It is a tough job. The biggest problem is the man-hours per flying hour, which is 40. I have 186 maintainers working on the jets." With 170 personnel, a Hornet squadron has slightly less manpower available for a plane that needs much less maintenance efforts. "With the Tomcat more things break and we have to work harder to get them flying."

Lieutenant Turner started in 1990 as an electrician working with Tomcats. "When I started in the Navy they warned me not to go to F-14s because of the hard work. But it is a very rewarding job." On New Years' Eve 2002 he got his orders for VF-31 and was on his way to the ship in a few days time. Despite the hard work and long working days he still is happy to be with VF-31. "It is very special to be in this very last cruise because the F-14 was the first plane I worked on."

Commander Rick 'Twig' LaBranche, commanding officer of VF-31, praises his maintainers. "Most maintainers do more than 12 hours per day," he maintains. "They want to fix the problems and really do a great job, they are a very proud bunch of people." While working on a Tomcat in

Over Iraq the main task is to provide on-call close air support against targets that may be encountered by ground forces. Imagery from the Tomcat's LANTIRN pod can be transmitted to ground-based laptops by the ROVER datalink.

the hangar bay AME (Aviation Structural Mechanic Safety Equipment) Richard Baxley of VF-31 does not complain about his long working days: "Because of the busy long days, the cruise goes faster." His colleague, AME Justin Jaehnig, adds that it is just like one's first car. "You put your blood, tears and sweat in it but you don't want to get rid of it."

According to Commander LaBranche, the avionics are prone to problems in particular, but the problems are not all about avionics as Lieutenant Junior Grade (LTJG) Rip 'Flounder' Gordon experienced during an ACM training hop in the Gulf. "I was flying FELIX 104 when the flaps of one of my wings inadvertently went down. It happens sometimes during ACM. The result was that the starboard side of the fuselage behind the wings was damaged and I had to divert to Kuwait. The landing was a little tricky but I had first made some practice landings at altitude so I was kind of prepared." According to Commander LaBranche it takes four guys one day to fix this flap lock problem

Parts

Commander LaBranche contends that there are no real maintenance surprises. "The secret is that we look ahead, try to avoid problems in the future instead of waiting for them to happen. For instance, when we have not had a certain engine problem in the last two months we say 'let's go check them'. We also communicate intensely with our maintainers to explain any problems." Spare parts are not a problem according to LaBranche. "Globalisation means a smaller world so we have our parts from the US in a few days. We also don't get surprised anymore. We have done this for so long, we know an awful lot about the plane."

As there are no more Tomcats available elsewhere it is not possible anymore to replace a broken or downed jet but to Commander LaBranche that is not really a problem: "We just

A 'Black Lions' Tomcat wheels high over the northern Iraqi city of Mosul. Tomcat crew works closely with forward air controllers on the ground, making the two-seat Tomcat ideal for such missions.

bring in another Hornet instead."

Proudly, VF-31 Tomcats wear the symbol of the 2005 Commander, Naval Air Forces, Atlantic Fleet (CNAL) Phoenix maintenance award on the tail. This award is presented to the best performing maintenance department of all east coast US Navy and Marine Corps squadrons. Being rewarded with this award is a great accomplishment because of the Tomcat's very high maintenance demand, and is a big morale boost for the maintainers.

Nugget impressions

A number of young aviators had the pleasure to be on their first cruise. Among these so-called 'nuggets' are the pilots who graduated in spring 2005 in the last ever Tomcat class at NAS Oceana with VF-101.

LTJG Daniel 'Bunny' O'Hara, who made the graduation barely, is now a comfortable well-performing pilot in VF-31. He got assigned to VF-31 on 22 March and instantly found himself in a tense and busy work-up schedule leading to the 2005-2006 cruise. "When I arrived I immediately got back to the boat for a month. Sometime later we went to NAS Fallon for advanced carrier air wing strike training. I dropped real bombs for the first time and participated in big and complex strikes involving many planes. It was awesome to see the whole air wing working together and be part of it." At the Fallon work-ups he also fought the Navy aggressors and even bagged an F-16.

For every new pilot in a squadron he first has to prove his abilities in the air and 'Bunny' was no exception. A young pilot fresh from the RAG has only started to learn his skills but he has also to perform. "You feel like you're being observed under a microscope. Everybody is watching you. But I knew what to expect because I had the same experience in VF-101 when I arrived. Still, I was pretty nervous and initially my landings grades could have been better, but they have improved

now. Getting confident took a while. A lot of guys are looking after me to make sure I do it safe."

LTJG Matthew Nieswand had similar experiences. "I had to make a name for myself in the squadron. I am always the newest guy," and, he counters with a sore grin when a fellow squadron member chimed in with some unwanted, and likely undeserved, comment on his performances "I have nobody to pick on." LTJGs Gordon, Nieswand and O'Hara are all performing well. LTJG Gordon even received the CVW-8 Top Nugget Hook award for a certain line period (two months). Flying from aircraft-carriers is in his genes. His father, Commander Roy 'Flash' Gordon, flew F-4s and F-14s and was the skipper of VF-31 during Desert Storm.

LTJGs Gordon and O'Hara made a significant contribution to the prestigious tailhook award VF-31 earned in the summer of 2005. Even the CAG was surprised the 'Tomcatters' won this award because the Tomcat is not known for its pretty handling characteristics in the landing pattern. In his career CAG Sizemore never saw a Tomcat squadron earn it. LTJGs O'Hara and Nieswand are now in LSO training. Only pilots

with good landing performance are asked for this highly respected job.

'Rhino' transition

In late summer 2006 VF-31 transitions to the F/A-18E Super Hornet and becomes VFA-31. The original plan was a conversion to the two-seat F/A-18F but, according to Commander LaBranche, it is a matter of timing. "An important reason is logistics. If we had to transition to the F/A-18F we had to go to Lemoore, California. Now we stay in Oceana, Virginia, so that we do not have to move a lot of people to the West Coast."

VFA-22 will exchange its F/A-18Es for the F-models that were initially planned for VF-31. VFA-22 reports to Carrier Air Wing 14. The second Super Hornet squadron is VFA-115 that is also equipped with the E-model. The US Navy wants one E- and one F-squadron in every air wing so, with this swap, that problem is solved for

The aircraft assigned to the 'Tomcatters' commander wears the traditional red markings with black radome. The 'Felix' badge has been used by VF-31 since the unit was created by the renumbering of VF-3 in 1948.

CVW-14. After the transition VFA-31 stays in CVW-8 alongside VFA-213 that transitioned from the Tomcat immediately after this cruise.

Undoubtedly the change to the E-model and the consequential stay in Oceana was also decided because of the busy deployment schedule VF-31 had in the last three years. In May 2002 the 'Tomcatters' went on a cruise with CVW-14 in *Abraham Lincoln* (CVN 72) for a supposed regular six-month cruise. However, this cruise became the longest in 30 years as it lasted until July 2003 and covered all of the combat phase of Operation Iraqi Freedom. In May 2004 the squadron was back on the boat with CVW-14 but this time it was the USS *John C. Stennis* (CVN 74). It returned from this six-month cruise in late October 2004. This was the last WestPac Tomcat cruise.

Because of the transition schedule of the other Tomcat squadrons that flew the older A and B models, the US Navy ordered VF-31 to move to AIRLANT: from Pacific Fleet CVW-14 to Atlantic Fleet CVW-8. Almost immediately it started the busy 10-month work-up for this final Tomcat cruise. The schedule is burdensome for the crew and their families.

Return to sea

The high spirits and performances of the 'Tomcatters' is all the more astonishing when one realises that normally a squadron goes on a regular cruise about two years after their last. Commander LaBranche underlines the fact that it was very difficult for his people to go on a cruise again. "They are loyal to the squadron, to the

Not as colourful as '213', VF-213's 'CAG' and 'boss' (illustrated) jets have high-visibility markings and an extended anti-glare panel forward of the cockpit. The LANTIRN pod is always carried on the starboard pylon.

Navy and to their country. They understand the importance of the mission. When you work five days, get home tired and see your neighbour's home is on fire you don't do nothing but go to help. They really do a great job." He also emphasises that the Navy leadership does a great job. "They provide great support to us and our families at home. Without that it would have been very difficult for us all."

Commander LaBranche made the transition from the Grumman A-6 Intruder to the F-14D in 1994. The D was just entering service at that time while the Intruder was being retired. "I was lucky.

GBU-38 – the little JDAM

The weapons of choice for regular operations over Iraq are based on the 500-lb (227-kg) Mk 82 bomb. The GBU-12 laser-guided bomb has been part of the Tomcat's arsenal since the LANTIRN pod was added to the wing station. The GBU-38 500-lb version of the GPS-guided JDAM bomb is a different story, because it was only cleared for employment by the Tomcat a couple of days before the final cruise got under way.

The Tomcat put the JDAM in use for the first time in February 2002 in Afghanistan. This was the heavy 2,000-lb GBU-32 version of this weapon. However, the explosive size of this weapon limits flexibility against the kind of target encountered nowadays in Iraq. The 500-lb variant on the contrary minimises collateral damage. Tomcats carry both the laser and GPS-guided weapons on the same mission. The laser-guided weapon has certain advantages over the GPS-guided weapon, and vice versa. The latter, for instance, is an all-weather bomb that can be used in adverse weather and sandstorms, while the laser weapon is more accurate.

During their pre-cruise work-ups VF-31 and VF-213 knew what kind of weapons would be preferred over Iraq and realised that they needed the small JDAM bomb in order to remain a primary player. They approached the programme office in March 2005, asking whether it could clear the aircraft to use the 500-pounder.

The test communities at Naval Air Station Patuxent River and Naval Air Warfare Center Weapons Division (NAWCWD) China Lake retired their Tomcats in 2000 and 2004, respectively, but there were some test pilots with Tomcat experience who could do the tests. They were more than happy to get back in the cockpit of the 'Big Cat' again. They performed the initial tests that included behaviour of the weapon attached to the bomb rack under extreme conditions such as dives, high-speed manoeuvring and carrier landings, followed by weapon separation tests.

Crews from VF-101 'Grim Reapers' took over for the guided weapons release tests at Naval Air Station China Lake, California. VF-101, the Fleet Replenishment Squadron, was at that time finished with training new crews and had only a few instructor pilots and RIOs left as this squadron was preparing for decommissioning. From day one in the northern Arabian Gulf the F-14s of VF-31 and VF-213 had GBU-38s in the air. The standard payload in October 2005 comprised one AIM-7 Sparrow or AIM-9 Sidewinder air-to-air missile, and one GBU-38 and GBU-12 bomb between the engine nacelles.

A GBU-38 (left) is carried next to a GBU-12 (right) on this VF-31 F-14D, representing the standard air-to-ground load for Tomcats on Iraq policing duties. Bombs are carried under the pallets previously used for carriage of the AIM-54 Phoenix long-range AAM.

In order to remain flying I had to make a transition to another aircraft. It was a great deal because the D was brand new then. The technology of the D is very good. It has a very strong radar, the strongest available. It is a very capable aircraft with 60,000 pounds of thrust." Though he loves the Tomcat he looks forward to the Super Hornet, nicknamed the 'Rhino', because of all the new technology involved. "I am confident Navy lead-

Although capable, the F-14D is labour-intensive compared to its Hornet deck-mates. The armament trolleys carry GBU-12s, GBU-38s and AIM-7 Sparrows, although this VF-31 aircraft is armed with an AIM-9.

ership is giving it all the next-generation technology. The 'Rhino' will become what the F-14 became: the best air-to-ground platform in the Navy."

As the cruise passes the transition to the Super Hornet is becoming more and more in the mind of the crews, especially for the three pilots that made up the last Tomcat class to qualify on the type in spring 2005. "It is strange learning a jet while it is getting away," reflects LTJG Matt Nieswand of VF-213, "and it is particularly strange to transition to a new plane again immediately after this cruise." Nieswand has the distinct

VF-213 commander's aircraft flies over western Iraq in mid-November 2005. At the time the Tomcats were supporting Operation Steel Curtain against al-Qaeda insurgents entering Iraq from neighbouring Syria.

honour of being the very last Tomcat pilot to qualify. LTJG O'Hara made the qualification in the Tomcat just barely. He is excited to make the transition though with mixed feelings. "I finally get a good idea of how to use the F-14 but it is only for a very short period. That is kind of frustrating, but at least I can say I have flown the Tomcat. It will be strange because I have not flown a single-seater in a very long time. I will have to do everything,

Although the Tomcat has impressive range, inflight refuelling is an integral part of policing missions over Iraq to extend time on station. Here Air Wing Eight Tomcats refuel from a KC-135R (above and below) and an S-3B Viking (right). The Viking provides the air wing with its own refuelling capability, but as the S-3s are retired so the task is being entrusted to the Super Hornet.

such as radio and radar, all by myself again."

Like LTJG Gordon, and many other aviators, LTJG Scott 'Timmey' Timmester followed in the footsteps of his father, who flew three tours in F-4s (including Operation Linebacker II in Vietnam) and one in F-14s as a radar intercept officer (RIO). Before he entered training as a Naval Flight Officer, 'Timmey' encountered some setbacks on his way to a Tomcat cockpit. Initially, his eyes were not good enough to be a RIO, let alone a pilot, and had to focus on a civilian job instead. "I had a boring job in computer programming consulting when one day a friend told me that the physical limits for RIOs had been lowered. A dream came true. You know, I really wanted to be in Tomcats. In junior high school I used to draw F-14s all the time. I knew all the Navy squadrons and their tailcodes. So, within four days I quit my job and used six months to get in the shape for the tough physical examination."

By mid-October 'Timmey' had flown 230 hours in the Tomcat and he loves being a RIO. "As a RIO you can be very busy doing a lot of 'admin', talking to people on the ground, but there is also time for sightseeing. One day we showed up just after an IED went off and I saw troops secure the area. It was awesome to look at it through our LANTIRN pod."

His father did not really want to make the transition from his beloved Phantom to the F-14. This is a common phenomenon when aviators transition from a well-performing platform to its unproven successor. Many F-14 aviators are proud of their plane and do not really want to make the transition to the Super Hornet.

For 'Timmey' the Rhino transition will be bittersweet. "I have wanted to fly the Tomcat my entire life, so I will be sad to see it go. Nonetheless, it will be great to fly a jet that always works, and that doesn't require an exorbitant amount of work by maintainers to keep it flying. Plus the new avionics such as AESA, Joint Helmet Mounted Cueing System [JMHCS] and weapons such as AMRAAM should make up for the hit we'll take in airspeed. From what I hear, the situational awareness will be massive for the aircrew of an AESA-equipped Super Hornet, and I'm definitely looking forward to that."

3,000 hours

In the more than 10 years he flew the Tomcat Commander Rick LaBranche logged more than 2,000 Tomcat hours (in addition to 1,000 in the A-6). Commander Howe has been flying Tomcats since he entered type conversion at the end of the 1980s and had in October 2005 some 2,550 hours on the counter. His goal is to make 3,000. "I hope to become the very last Tomcat aviator to make 3,000 hours. It is going to be close. I have still 450 hours to go before we enter Super Hornet transition. If I make it, it will be in the very last month," he calculates. "We have still 4 and a half months to go on this cruise and I fly approximately 60-70 hours per month."

The high hours are partly caused by the long missions flown over Iraq in support of Operation Iraqi Freedom. "It hurts your rear end," comments pilot LTJG Gordon on the six- to eight-hour missions. LTJG Scott 'Timmey' Timmester adds that they can be very boring. These missions take the Tomcat far into Iraq and require three aerial refuellings. Usually, the Tomcat aviators fly these missions every three days.

Deck crew swarm over two Tomcats from VF-31 and a VF-213 machine to prepare them for the next mission. The F-14s are up to the heavy tasking schedule but only through the application of many man-hours.

The USAF supplies much of the refuelling for Navy aircraft over Iraq, and here Air Wing Eight aircraft take on fuel from a Travis-based KC-10A Extender. US Navy fighters also occasionally refuel from RAF tankers.

LTJG Gordon and a number of other guys have now flown their first combat sorties, although they emphasise that, although the OIF missions are indicated as combat, there are hardly, if any, enemy air defences aimed at them.

In the beginning, especially, it was pretty exciting for them. "It was fascinating to go on my first combat mission," recalls LTJG Gordon. "I always watched it on television but now I saw it myself. First it was exciting but now it is benign."

For the RIO there is a lot more going on. While the pilot is flying the aircraft the RIO is doing the targeting and communications with the people on the ground. For them it is much more exciting as LTJG Timmester acknowledges. "We are in contact with the people on the ground and we talk about threats and 'possible' targets." 'Timmey' is not FAC-A qualified but that, he explains, is only required under certain conditions such as heavy close air support or when there is no FAC on the ground. CAS is the application of combat aircraft against ground targets in close proximity of people you don't want to hit.

The technology has advanced in such a way that target identification has become much less demanding. "The Joint Terminal Attack Controller (JTAC) on the ground has certain tools to give you precise target coordinates. Our targeting pod cues automatically towards it. We verify it with the JTAC to be sure we are looking at the same target. Then I talk the pilot to it. At night we see through our night vision goggles the infrared marker of the JTAC pointing at the target. We also see the infra-red markers of our own troops so the

chance of friendly fire is reduced."

Verification with the JTAC was simplified in mid-December when the Remotely Operated Video Enhanced Receiver (ROVER) was introduced in the F-14. With ROVER no more time-consuming vocal verification is needed. This system allows transmission of real-time images acquired by the Tomcat's sensors to the laptop computer of the JTAC. It makes it a lot easier for him or her to determine if the target is real and allow it to be targeted. It is also a great reconnaissance asset for people on the ground as they see the area of operations from a bird's eye-view.

Combat ops

The level of aggression put into force by the Tomcats depends on the nature of the (possible) threat. According to Timmester, a 'show of force' by flying fast and low, sometimes at 500 ft (152 m),

is part of the tactics, while dropping bombs is the other end of the spectrum. In the first months of OIF operations VF-31 and 213 Tomcats were involved in several such actions.

On 15 October 2005, two F-14s performed pre-planned strikes against an insurgent weapons cache in the vicinity of Ar Ramada. This attack came shortly after two CVW-8 F/A-18Cs provided close air support to coalition troops in the vicinity of Karabilah. The targets were buildings being used by anti-Iraqi forces as firing positions. On 19 October a F-14 of VF-31 destroyed an Improvised Explosive Device production weapons facility northeast of Baghdad.

Transformed from a sleek 'Cat' to an ungainly 'Turkey', an F-14D brings a GBU-38 back to Roosevelt *after an Iraq mission. Note the open doors in the intakes that alter the capture area.*

Carrier Air Wing Eight

The air wing embarked on the USS *Roosevelt* is CVW-8. Six squadrons escorted the two Tomcat squadrons on their last cruise. Two F/A-18C Hornet fighter-bomber squadrons (VFA-15 'Valions' and VFA-87 'Golden Warriors') provided air-to-ground capabilities. VS-24 'Scouts' made their last cruise before disestablishment in September 2006. They performed vital aerial refuelling for the air wing and conducted Maritime Security Operations with S-3B Vikings. VAQ-141 'Shadowhawks' employed its four EA-6B Prowlers from the *Roosevelt*, as well as shore bases. Prowlers are key players when it comes to jamming the radios, telephones and other communication channels – including those controlling IED devices – used by the insurgents. HS-3 'Tridents performed plane-guard duties as well as a host of support missions in the Gulf and Iraq with their HH-60H and SH-60F Seahawks. The E-2Cs of VAW-124 'Bear Aces', in NP2000 configuration, were the ears and eyes in the air.

Above: A sight never to be repeated: Tomcats cluster around the stern of Roosevelt *as the carrier sails in the Arabian Gulf. Operations concentrate on providing support to forces in Iraq, as well as maritime surveillance in the northern gulf. US Navy Middle East operations are supported by a headquarters in Bahrain.*

Partnering the Tomcats in flying armed reconnaissance and support missions over Iraq were two squadrons of F/A-18Cs: VFA-15 (below left) and VFA-87 (CAG-bird below). The Hornets typically fly with a pair of GBU-38 JDAMs, plus a single AIM-9X on the wingtip to give a measure of self defence.

Operation Steel Curtain was an offensive that started on 4 November that was aimed at preventing cells of al-Qaeda from entering Iraq through the Syrian border and restoring Iraqi sovereign control along the Iraq-Syria border. It also aimed to destroy al-Qaeda terrorists operating throughout the Al Qa'im region. Involved were 1,000 Iraqi Army soldiers and 2,500 US Marines, who began the offensive near the town of Husaybah near the Iraq/Syria border. Military officials said Husaybah had become a haven for cells of al-Qaeda entering the country from Syria. At that time it was the largest military assault since American-led forces stormed Fallujah.

CVW-8 began providing air support for this operation on 6 November with both reconnaissance and strike missions in support of the ground troops. VF-213 and VF-31, along with Hornets, conducted several strikes on locations being used by Anti-Iraqi Force (AIF) personnel for strategic firing positions against US Marines and coalition ground forces.

In addition to flying missions over Iraq, *Theodore Roosevelt*-based aircraft have been flying missions in support of Maritime Security Operations in the Persian Gulf. US Navy sources say that these missions set the conditions for security and stability in the maritime environment, as

well as complement the counter-terrorism and security efforts of regional nations by denying international terrorists use of the maritime environment as a route for attack or to transport personnel, weapons or other material. MSO missions usually take around 90 minutes, as the north Arabian Gulf is not that big an area to cover. Other missions flown from the carrier are convoy escort, oil pipeline patrols and the protection of new construction.

The first Tomcat cruise took the swing-wing jets into hostile action when they provided air cover in 1975 over Saigon, Vietnam, during the evacuation of the US Embassy. Now, more than 30 years later, the type bows out in combat too as the spearhead of US diplomacy. It is sad, as the Tomcat is not an old and ageing weapon platform urgently in need of replacement. On the contrary, it is a jet still in excellent shape and still fit for the most difficult missions.

Gert Krombout

Left: VF-31's 'CAG-bird' has black fins with the squadron badge worn large. On the inside of the fin is the Air Wing Eight badge.

Right: The Tomcat was originally scheduled to leave service around 2008, but the date was brought forward considerably as Super Hornets flowed from the production line at St Louis.

Air Wing Eight support: above is an S-3B from VS-24. Like the Tomcat, the Viking is fast-disapperaing from US Navy carrier decks. Right is an SH-60F from HS-3, while below is an E-2C of VAW-124. The Hawkeyes have been refitted with eight-bladed propellers wlth curved blades

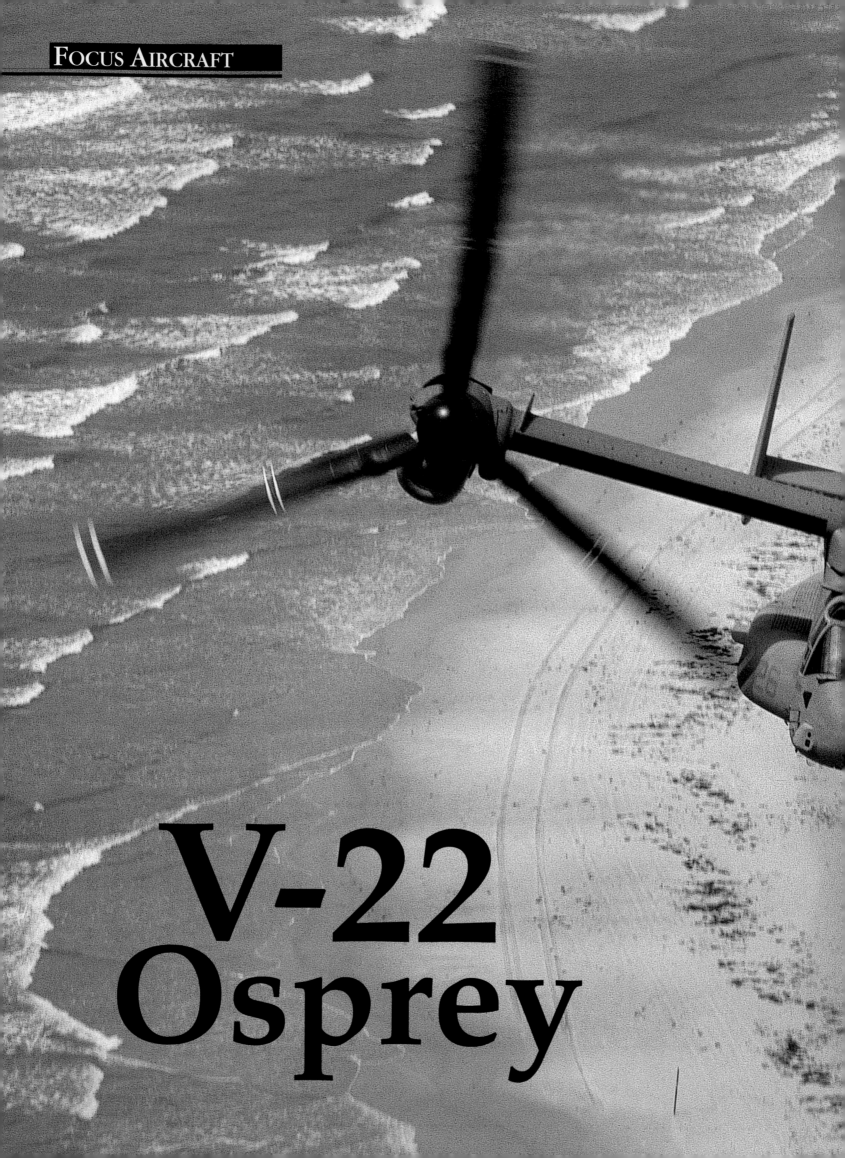

V-22
Osprey

The tiltrotor
comes of age

The year 2005 saw the Bell-Boeing V-22 Osprey programme reach three huge milestones, namely a successful Operational Evaluation (OPEVAL) completed during the summer and approval for full-rate production in late September, followed by the first delivery of the MV-22B Block B configuration in December. Marking a historic point in a programme that began in the early 1980s, the Marine Corps is now poised to begin integration of the Osprey into combat units, beginning with VMM-263, the first operational MV-22 squadron, which stood up on 3 March 2006. VMM-263 received its first Osprey, aircraft No. 73, in April.

The MV-22B Osprey is replacing the CH-46E and CH-53D helicopters used by the Marine Corps, and the CV-22B will replace MH-53J Pave Low helicopters used by Air Force Special Forces. The Marine Corps plans to purchase 360 MV-22s and the Air Force Special Operations Command (AFSOC) to buy 50 CV-22s. The Navy also has an unfunded requirement for 48 MV-22s for possible use in fleet logistical support and search and rescue, although present budgetary constraints seem to suggest these may never come to fruition. Initial operational capability (IOC) is scheduled for 2007 for the MV-22 and in 2009 for the CV-22.

International Air Power Review examines the Osprey programme since its 2002 return to flight status, highlighting the most recently completed OPEVAL and providing an overview of the current and planned configurations. The Osprey introduces capabilities currently unavailable in any other rotor platform and presents a significant capability to American armed forces. The Osprey is twice as fast as a helicopter and can carry three times more payload, or cover five times the range of the CH-46. This additional range not only increases the distance from which the V-22 can launch, but also expands the battlespace, thereby complicating any adversary's defensive solutions, as the Osprey can adjust its flight plan to avoid threats. In remarks made last December, Lt Gen. Jim Amos, Commanding General, II Marine Expeditionary Forces, summed up exactly what the Osprey brings to the Marines capabilities: "The Osprey remains at the very soul of our Corps' ability to fight future conflicts across a widely dispersed battlefield. Battlefields where the tyranny of distance is solved with speed, and where an irregular enemy who chooses to fight at an urban marketplace or at an ambush site in a wadi is faced with the dilemma – 'where are they? I know they are coming, I just don't know when or where.'"

Return to flight – 2002

May 2002 marked the return to flight for the Osprey programme following a nearly 18-month operational pause. During that time, two panels reviewed the Osprey programme, both concluding that it offered significant value and recommended further development. On 19 April 2001, an independent Blue Ribbon panel issued its report to the effect that there was nothing inherently wrong with the tiltrotor technology.

According to the report, starting over with a new programme would take years to develop and many millions of dollars to get where the V-22 had already reached, and in all likelihood, with lesser capabilities. In going forward with the Osprey, the panel suggested additional training and testing to address the Vortex Ring State (VRS) problem (discussed later),

A NASA panel further examined the tiltrotor aeromechanics and reported in November of that year in essentially the same manner as had the Blue Ribbon panel. It, too, called for more testing of the VRS problem and urged the elimination of the requirement for auto-rotation to a survivable landing under all conditions. Other reports followed, most notably that of the Defense Science Board, which suggested that too many outside pressures had forced programme managers to cut corners in testing and development. It was recommended that the programme be restructured from a date- to event-driven schedule.

VRS phenomenon

Vortex Ring State, referenced in several of the reports, is certainly not an uncommon phenomenon in rotor aircraft. VRS occurs when a helicopter descends at a high rate of speed but with little forward velocity. Since the rotor is descending through its own downwash, the rotor blades are subject to

Above: An Osprey from the test fleet flies near the Navy's main test centre at Patuxent River. The rear ramp allows for rapid loading and unloading.

Below: Wearing 'Navair Team Osprey' titles on the fin, this MV-22B of the integrated test team is seen aboard USS Iwo Jima during shipboard trials.

Above: VMX-22 was established in August 2003 at New River with the main aim of conducting the OPEVAL. It continues in the test and evaluation role with a reduced complement.

Above: A key requirement for the Osprey was that it could be folded to fit into a tight space. Here the reverse process is two-thirds complete, with the wing swivelled to its inflight position and the engine nacelles swivelled to the vertical. The rotor blades are about to be unfurled. This VMX-22 aircraft was on USS Wasp at the time.

Right: The helicopter's ability to carry underslung loads has not been lost with the Osprey. Here an ITT MV-22 carries a 'Hummer', a typical load. The cabin floor has a 'hell-hole' through which the load can easily be monitored. Underslung loads can only be carried in the pure helicopter and conversion mode.

stalls and resulting loss of lift. VRS is believed to have contributed to the loss of aircraft 14 on 8 April 2000 during an OPEVAL exercise. Studies during the operational pause led to incorporation of new software to help pilots identify VRS conditions (sink rate indicators and an audible voice saying

'sink rate') and additional training measures, as well as validation of the flight envelope.

Rather than modifying the flight envelope, as some have suggested occurred, the programme conducted the most extensive flight test programme of any rotary-wing aircraft to validate the VRS onset envelope. The aircraft was purposely flown up to and through the onset of VRS, with the following conclusions drawn: (1) even with conservative safety limits, the Osprey can descend as fast or faster than any other aircraft we operate; (2) although the programme set a safety limit of 800 ft (244 m) per minute max descent rate under 40 kt (74 km/h) – the standard for all DoD rotary-wing aircraft – the early signs of VRS did not manifest until 1,600 ft (488 m) per minute and full-blown VRS departure does not occur until above 2,000 ft (610 m) per minute; (3) during the Marana mishap, the aircraft was descending at approximately 2,200 ft (670 m) per minute; and (4) unlike any other aircraft, the Osprey can get out of VRS simply by moving the nacelles forward by 16° (which takes about two seconds) and powering into clean air.

Low-rate production

Despite the operational pause, the Osprey programme continued production, although at a low rate. According to Mike Anderson, Bell-Boeing V-22 Joint Program Office Vice President and Deputy Program Manager, continuing production, even at the lowest practical rate, meant the retention of manufacturing knowledge and skilled labour key to the programme, and further kept contractors and vendors committed to the overall effort. It was understood that these production aircraft, which were temporarily placed in storage, would be modified later to incorporate suggested changes and bring the aircraft to the Block B configuration.

Indeed, Bell-Boeing continued development of the aircraft, incorporating changes and recommendations made by the Blue Ribbon panel established to evaluate the programme following the 2000 crashes and by NASA. A list of several hundred changes was compiled and assessed, then prioritised with an eye to obtaining an aircraft capable of return to flight. Of these items assessed, only a fraction had to be corrected and confirmed with successful flight tests, before any of the training aircraft could return to flight. These changes were dubbed Block A and consisted of largely nacelle redesign and software upgrades.

Above and below: The V-22 has a thoroughly modern cockpit, each pilot having two large display screens. A large shared screen in the centre (with threat warning display to the right) usually displays aircraft systems data such as engine readings. Flight control is effected via the centrally mounted cyclic control stick and the left-hand Thrust Control Lever. A thumbwheel on the TCL allows the pilot to change engine nacelle angle. An automatic nacelle angle function based on airspeed is under development.

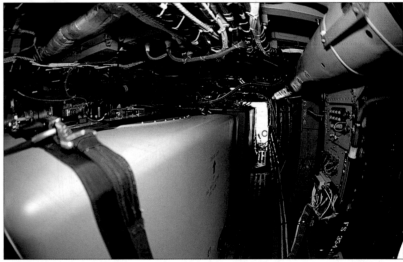

Above: The cabin is sized to accommodate 24 fully-equipped troops, but can also carry a sizeable cargo load. Here an HX-21 aircraft is fitted with three Robertson Aviation rigid tanks that give the Osprey a 2,100-nm (3888-km) unrefuelled self-deployment capability.

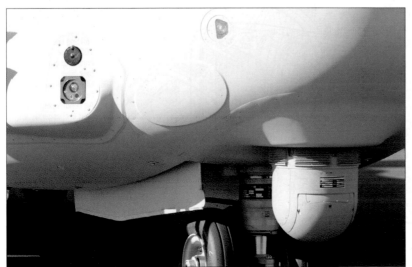

Left: Below the nose of the MV-22B (and CV-22B) is a turret-mounted AAQ-27 MWIR (Mid-Wavelength Infra Red) imaging system. Also visible is the blanking plate covering the position of the refuelling probe, when fitted.

Right: On either side of the nose are antennas for the APR-39A radar warning sensor (upper) and AAR-47 missile warning sensor (lower). The CV-22B differs by having the ALQ-211 SIRFC in place of the APR-39.

Above: Navy SEAL (Sea-Air-Land) special forces parachute from the open rear ramp of an ITT MV-22.

Above right: A group of Marines boards an Osprey at Camp Lejeune. The position of the rotors means that entry to the Osprey's cabin is easier and safer than for comparable helicopter types.

Below: This VMX-22 MV-22 is parked in the fully-folded configuration. The folding and unfolding processes each take about 90 seconds to perform. Note the fixed refuelling probe on the aircraft in the background.

The remaining changes were divided into Block B and C, the former relating to enhanced suitability, improved capabilities, and nacelle maintenance access, and the latter dealing with improvements to mission capabilities. All changes were to be incorporated as modifications to existing aircraft or introduced into the production line as they became available and funded.

On 29 May 2002 aircraft 10 resumed flight testing operations at Pax River. Aircraft 8 followed on 19 October. EMD testing, which had initially begun in 1992, resumed using five fleet-representative aircraft, Nos 21, 22, 23, 24 and 34. Aircraft 21 constituted the first production aircraft to be modified with Block A changes and number 34 marked the first production MV-22B Block A aircraft, arriving on 20 August 2003. This aircraft incorporated various software modifications, weight reduction initiatives, and featured a redesigned nacelle and safety of flight features.

EMD testing continued during 2003, adding another 1,000 hours of flight testing. Overall, the 2003 tests were very productive and without any major incidents. Testing focused on a re-examination of combat manoeuvring, aerial refuelling,

and dust and debris re-circulation during landing, as well as two at-sea periods to examine the uncommanded roll-on-deck phenomenon that had been experienced when one Osprey hovered in proximity to another on deck. That testing resulted in specialised control laws that effectively mitigate the roll-on-deck phenomenon.

Test fleet deployment

Of the test aircraft, No. 10 deployed to the USS *Iwo Jima* in January and was joined by No. 22 later in November aboard USS *Bataan*. Aircraft 10 also conducted additional aerial refuelling tests and joined No. 22 on various mission system tests. Aircraft 21 handled night formation work while Nos. 21 and 22 conducted austere landing trials. No. 21 also deployed to Fort Bragg for development of air-drop capabilities, where tests evaluated parachute-recovered loads of up to 2,000 lb (907 kg) and paratroop drops. Aircraft 24 was fitted with a simulated CV-22 radar for icing testing in Nova Scotia during December 2003 through April 2004. This aircraft, dubbed 'Chilly Willy' by the integrated test team, has returned to Nova Scotia (Halifax) to obtain additional environmental data points in the winters of 2004/2005 and 2005/2006.

Flight testing picked up its pace during 2004, with aircraft 9, now converted to the CV-22B configuration, flying open-air range electronic warfare flights at China Lake's combat range in March. In April, the V-22 ITT resumed air-to-air refuelling flights for the first time since the return to flight in May 2002. The ITT crews consisting of Lt Col Kevin Gross and Boeing test pilot Steve Grohsmeyer each connected with the refuelling aircraft five times near Pax River, although no fuel was actually passed. Aircraft No. 22 was outfitted with an 11-ft (3.35-m) fixed refuelling probe. A new retractable refuelling probe, measuring 9 ft (2.74 m) when deployed, was installed on aircraft 21. The retractable boom is crucial to the Osprey's ability to perform shipboard operations. April 2004 also marked 1,330 flight hours since return to flight and further saw the

The use of simulators is an increasingly important part of military training, and perhaps more so for the Osprey as it combines the characteristics of both fixed- and rotary-wing craft.

Navy drop the HV-22 designation in favour of the MV-22 designation used by the Marines variant.

Icing testing was conducted in April at Canadian Forces Base Shearwater near Nova Scotia, where aircraft 24 logged 67 flight hours, 37 of which were in icing conditions. Additional ice testing was conducted later in 2004 and in 2005 at Halifax (as referenced above).

On 18 May, a seven-week pre-OPEVAL assessment began to evaluate the impact of the Block A modifications, which resulted from the prior OPEVAL and other studies, and to identify risk areas for the OPEVAL as well as supporting continued development.

In June, the ITT completed Phase IVB of the shipboard suitability testing (the fifth of six at-sea periods), with aircraft spending eight days aboard the USS *Iwo Jima* (LHD 7) off the coast of Maryland. This testing served as a follow-on to 1999 shipboard testing, which had revealed a tendency for the Osprey to tilt along its lateral axis when sitting on the flight desk behind a hovering MV-22B aircraft. Reprogrammed flight controls ended this uncommanded roll, and the testing was deemed successful.

MCAS Yuma in Arizona is an important Marine base and a regular destination for exercises and trials. Here an MV-22B from HMX-1 practises in the desert prior to the first of two crashes that led to the fleet grounding in 2000.

As the programme moved into August and September, it surpassed the 3,000 flight hour mark since return to flight. At that time there were nine test aircraft in the V-22 test flight programme. Two were at Edwards (the CV-22B aircraft 7 and 9) and seven were at Pax River. VMX-22, which had stood up at MCAS New River on 28 August 2003 to perform the OPEVAL Phase II testing, possessed 11 aircraft.

Above: In June 2005 USS Bataan (LHD 5) played host to eight VMX-22 Ospreys as part of the OPEVAL. Here the octet spools up ready for a mass launch, one aircraft being positioned aft of the island.

Above: Two MV-22Bs from VMX-22 undergo trials aboard USS Iwo Jima in July 2004. In the background is a CH-46 Sea Knight, the type that the Osprey is replacing.

The year concluded with the final phase of the shipboard suitability testing, designated Phase IVC, which involved 10 days at sea aboard the USS *Wasp* (LHD 1) beginning on 12 November using aircraft Nos 10, 21 and 23. This testing completed the interaction testing between a V-22 parked on the flight deck and another V-22 hovering in front of it, and further expanded the flight envelope for all port-side landing

spots aboard the LHD. Tests also developed a night short take-off envelope and evaluated the latest flight control software versions. Aircraft 22 was also present and completed all planned below-deck heavy maintenance evaluations.

In anticipation of the upcoming OPEVAL, pilots from VMX-22 qualified aboard the USS *Kearsarge* (LHD 3) from 7-13 December. Pilot training continued into early 2005, but all flights were stopped due to a bearing problem. Several aircraft experienced problems with bearings in the gearbox involving the flaking of the thin dense chrome on the input quills and improper torquing of a part within the gearbox. This, in turn, caused a chip light alert signal to illuminate, which prompted emergency landings. After replacing the bearings with non-chrome coating, flights resumed on 7 February, with aircraft No. 57, the last of the eight reserved for the pending OPEVAL.

By mid-spring 2005 the stage was set for completion of the OPEVAL programme. It should be recalled that the initial OPEVAL (OT-IIE), which ran from November 1999 through 22 July 2000, found the V-22 operationally suitable and effective for USMC land operations, but because of reliability problems and the need for additional testing of the Blade Fold Wing Stow (BFWS) system, found the aircraft unsuitable for sea deployments. The report, issued 13 October 2000, represented the culmination of 522 sorties and 805 flight hours, indicated that the MV-22 was superior to the CH-46E and CH-53D. Of interest, the Osprey had been rated as operationally suitable by the Commander, Operation Test Force (COMOPTEVFOR), but the Director, Operational Test and Evaluation (DOT&E) withheld the "operationally suitable" rating due to outstanding issues.

OPEVAL deployments

The initial OPEVAL was performed at various locations, including Naval Air Station Patuxent River and the Naval Air Warfare Center Weapons Division in China Lake, as well as Marine Corps bases in North Carolina and Arizona, and Air Force bases in Florida and New Mexico. Sea evaluations were conducted aboard various amphibious ships (LHA, LHD, LSD) on each coast. Testing included self-deployment, land and ship-board operations, amphibious assault missions, over-water operations, night-vision goggles flights, low-level navigation, in-flight refuelling with a C-130 tanker, aerial delivery of cargo and personnel, austere landings, fast-roping, hoist operations, and external loads lifting on single and dual hooks. Various multi-aircraft flying formations were also evaluated.

Concerning the unsuitability finding, it is ironic that the programme was fully aware of this shortcoming going into the OPEVAL and, indeed, this recognition had justified many of the waivers imposed by Navy. In reality, the initial OPEVAL confirmed much of what the testing had already shown. The

For the Marine Corps the Osprey's main task is the OMFTS (Operational Maneuvers From The Sea) mission, which involves moving the assault element of the Marine Expeditionary Force from the ship to an inland objective. A squadron must be capable of completing this task in 90 minutes. As the aircraft is expected to spend much of its life at sea, a great deal of shipboard testing has been undertaken. Above a test machine is seen on Iwo Jima's elevator, while at right a pair of VMX-22 aircraft operate from Kearsarge in December 2004. Early sea trials revealed some problems with Ospreys operating in close proximity, requiring some control modifications.

Osprey met all key performance parameters and in many cases exceeded the programme's threshold requirements. The Osprey failed to meet only 23 of the 243 operational requirements, with the deficiencies considered minor and, in many cases, efforts were already underway to address the problems.

The Osprey programme fully intended to use the reliability data to target improvement efforts, which was apparent from its post-OPEVAL conduct. On 31 October 2000, an MV-22B deployed to the USS *Bataan* (LHD 5) for BFWS evaluations and passed with flying colours.

MV-22B OPEVAL – 2005

The Operational and Live Fire Test and Evaluation (OPEVAL) (also called the OT-IIG) recently completed in 2005 marked the second such test conducted on the V-22 Osprey. Its focus was to reassess the improved capabilities and recommendations incorporated into the Osprey from the two 2000 mishap investigations and the prior OPEVAL report. In short, the OPEVAL report found that the MV-22B Block A aircraft satisfied all key performance parameters and concluded that the aircraft was operationally effective, suitable and survivable. Here, it is worth noting that the Osprey that entered the 2005 OPEVAL was not the same Osprey that went through the initial OPEVAL. In fact, the changes to the aircraft were so extensive that the second OPEVAL more appropriately established the new aircraft's performance rather than build on the prior OPEVAL.

The OPEVAL process

The OPEVAL followed a rather detailed developmental test programme that ran from the 29 May 2002 return to flight through 31 December 2004. During this time, MV-22Bs flew 730 test flights with 1,433.1 flight hours and verified the correction of deficiencies identified during the prior OT & E

in 1999/2000 and subsequent reviews.

Conceptually, the OPEVAL has a different focus than EMD flights. EMD represents the period to test and confirm an aircraft's performance and test new parameters. The OPEVAL, on the other hand, tests the aircraft's ability to perform operationally and tactically in a realistic wartime environment and determines how it will fit into real-world operations. In essence, the OPEVAL report looks to answer, "How would this platform perform if I took it into battle tomorrow?"

Above: An MV-22B hovers over the sea. The Osprey can conduct the SAR role if required.

Below: The Marine assault ships of the future will have a full complement of MV-22s for the assault task, plus CH-53Ks for heavylift.

An important part of the OPEVAL was the evaluation of the Osprey in a variety of hostile environments, such as those that are currently being encountered by US forces in Afghanistan and Iraq where heat, dust and high altitude can be major factors. 'Brown-out' landing conditions are a feature of such operations, and were tested by the OPEVAL team at the Nellis AFB ranges in Nevada.

The goal of any programme heading through an OPEVAL is to receive the highest rating of "operationally effective" and "operationally suitable". In layman's terms, "operationally effective" means that the aircraft is able to perform its prescribed mission in a combat environment, and in the face of unexpected threats. "Operationally suitable" means that the aircraft, when operated and maintained by typical service personnel in the expected numbers and of the expected experience level, is supportable when deployed; specifically looking at how reliable the aircraft is and the adequacy of the supporting infrastructure.

The OPEVAL officially ran from 28 March to 29 June 2005, and used eight MV-22B Block A aircraft. The tests were conducted at several sites, namely MCAS New River, North Carolina, Mountain Warfare Training Center, Bridgeport, California, Nellis AFB, Nevada, with operations test ranges in California, Arizona, New Mexico and Texas. At-sea operations were conducted aboard the USS *Bataan* (LHD 5) in the western Atlantic with flights into ranges in North Carolina, Virginia, and Mississippi.

VMX-22 conducted the operational flights and accumulated more than 750 flight hours during 196 flight events during the three-month period. The flight hours included both events and functional check flights, which are required after significant maintenance activities. Of the 204 event flights, 89 were realistic end-to-end operational missions. The remaining flights, while not flown in a representative profile, provided data from which parameters could be extracted and compared.

OPEVAL II drew upon findings from three prior assessments, the Live Fire Test programme (LFT&E), which had been completed in 2000, and an Operational Assessment (OT-IIF), which ran from 18 May through 9 July 2004. The LFT&E conducted some 60 separate ballistic test studies (592 shells in all) with 7.62-mm armour-piercing incendiary (API),

OPEVAL confirms Key Performance Parameters (KKPs)

Profile	Demonstrated performance	Block A projection	Block B projection	Required threshold
Amphibious Pre-Assault/Raid (KPP)	230 nm (426 km)	247 nm (457 km)	247 nm (457 km)	200 nm (370 km)
Land Assault External Lift (KPP)	69 nm (128 km) with 9,980-lb (4527-kg) Lightweight Howitzer	63 nm (117 km) with 10,000 lb (4536 kg)	29 nm (54 km) with 10,000 lb (4536 kg)	50 nm (93 km) with 10,000 lb (4536 kg)
Amphibious External Lift	89 nm (165 km) with 6,900-lb (3130-kg) vehicle	115 nm (213 km) with 10,000 lb (4536 kg)	40 nm (74 km) with 10,000 lb (4536 kg)	50 nm (93 km) with 10,000 lb (4536 kg)
Land Assault Troop Lift	53 nm (98 km) x 2 trips	120 nm (222 km) x 2 trips	122 nm (226 km) x 2 trips	50 nm (93 km) x 2 trips
Self-Deploy (KPP)	Extrapolated to 2,660 nm (4926 km) with one refuel using Block B auxiliary tanks	n/a	2,400 nm (4444 km) with one refuel	2,100 nm (3889 km) with one refuel
Cruise airspeed (KPP)	255 kt (472 km/h)	255 kt (472 km/h)	250 kt (463 km/h)	240 kt (444 km/h)
Troop seating (KPP)	24 combat-loaded Marines	24 combat-loaded Marines	24 combat-loaded Marines	24 combat-loaded Marines

The MV-22B Block A accomplished all required mission profiles during the OT-IIG period. Block B's predicted drop in performance is attributable to increased weight in the aircraft, and was extrapolated using a very conservative model, factoring in the weight of all Block B kits installed at once.

12.7-mm API, 14.5-mm API, 23-mm API, and high-explosive incendiary (HEI), and 30-mm HEI. The latter test period assessed the impact of the Block A modifications on previously determined effectiveness and suitability, and identified areas of potential risk for the OT-IIG programmeme. VMX-22 also conducted an Austere Landing Deployment in September 2004, taking six aircraft to Nellis AFB to conduct flight operations into harsh landing environments. The team conducted 140 landings in five separate landing zones, presenting pilots with varying levels of reduced visibility ranging from light to severe.

Multi-phase testing

In evaluating the MV-22B's operational effectiveness, VMX-22 evaluated the Osprey's mission planning system and five specific mission types. The mission planning system was designed to allow aircrews to integrate charts and flight information assembled on a laptop or other mission system into the aircraft via a data cartridge. Crews could then modify or update the mission planning data while airborne and then at the conclusion of the flight transfer the information back to the cartridge for post-flight mission analysis.

The five missions evaluated included ship-to-ship manoeuvring, sustained shore operations, Tactical Recovery of Aircraft and Personnel (TRAP), self-deployment and amphibious evacuation. The Osprey's ability to manoeuvre defensively was also examined. The report found the Block A satisfied all key performance parameters (KPP) for these mission areas. The KPPs were established by the JVX Joint Operational Requirements Documents (JORD), which highlighted 13 criteria for MV-22 mission performance.

According to the report, there were very few limitations on testing. For example, VMX-22 conducted fewer operational night tests than planned. Pilots wore NVGs for only six percent of the total test time, only 12 of the planned 29 mission night profiles were accomplished (33 of the planned

131 flight hours). Twelve additional flights (totalling 14 flight hours) were flown for training and proficiency. This discrepancy was due in part to pilot currency issues, crew fatigue and ship availability problems. The test squadron also encountered fewer landings under severe visibility degradation or 'brownout' conditions, owing largely to a wet spring which created additional vegetation and minimised blowing sand and dust. Because of extensive 'brown-out' testing conducted in prior OT periods, however, the test authority had adequate data for this operational scenario.

These two photographs show an MV-22B from the Naval Air Warfare Center's test squadron HX-21 at Patuxent River. The engine nacelles swivel from fully forward (0°) to behind the vertical (97°). The latter setting allows the Osprey to hover backwards at up to 30 kt (56 km/h).

Above: Two of VMX-22's Ospreys display for the crowd at the 2003 New River air show, shortly after the squadron was formed. After the conclusion of the OPEVAL some of the unit's Block A aircraft were handed over to VMMT-204, the Osprey training squadron at New River.

Above right: The US Navy retains a healthy interest in the Osprey, and may use the type for COD, tanking and Combat SAR. The Navy's earlier HV-22 designation has now been dropped, and if the type is procured it will be known as the MV-22.

Below: Just as in a conventional helicopter, the crew chief keeps a good look-out from the side door as an MV-22B taxis. The engine nacelles effectively rule out the fitment of waist guns, but operational Block B Ospreys will have a gun mounted on the rear ramp.

VMX-22 issues its report

Operationally effective

Areas of Enhanced Features: The OPEVAL II report concluded that the V-22 system offered significant mission advantages compared to the CH-46 and CH-53 helicopters it will replace. These advantages include extended range with high speeds and larger payloads for greater operational reach and reduced response times, self-deployment capability (thereby reducing burdens on strategic lift), and advanced mission management systems and situational awareness equipment for precise navigation, increased battlefield situational awareness, and reduced aircrew workloads.

Areas of Concern Resolved from Previous Operational Testing: The report found that the Osprey programme had resolved four specific areas of concern from the first OT-IIE period and the safety problems associated with the two crashes in 2000. Of these areas, the two most important related to the VRS envelope and the impact of downwash on operations. The report noted that the VRS envelope was "now well defined and avoidable." VMX-22 crews were able to conduct all operational missions while remaining outside the VRS-susceptible envelope. Moreover, the impact of downwash had been minimised by revised tactics, techniques and procedures. Improvements were also noted in the flight control software reliability and hydraulic line routing.

Areas of Follow-On Testing Requirements: The report recommended three areas of follow-on testing and upgrades: a personnel hoist, defensive weapons system, and a

weather radar. Each of these were already incorporated as modifications in the various Block upgrade programmemes.

Operational suitability

The MV-22B satisfied operational thresholds for four critical areas:

■ Mean flight hours between aborts (25 hours versus > 17 hours required).
■ Mean flight hours between failures (1.4 hours versus > 0.9 hours required).
■ Maintenance man-hours per flight hour (7.2 hours versus < 20 hours required).
■ Mission capable rate (78 - 88 percent versus 82 percent required at 60,000 hours total fleet time).

The aircraft did not meet the threshold for mean repair time for aborts, but the team found this deficiency did not impact the overall operational suitability of the V-22. Abort-causing failures accounted for only 5 percent of the overall maintenance workload.

Concerning reliability, the VMX-22 report found the Osprey was a mature programme, as demonstrated by two key measures of reliability, mean flight hours between aborts (MFHBA) and mean flight hours between failures (logistics) (MFHBF). The Block A threshold requirement for MFHBA is 17 hours or greater and the Osprey achieved 25 hours. Of the

751.6 flight hours, the aircraft experienced only 30 aborts, either on the ground or in flight. The block A threshold requirement for MFHBF(log) is greater than or equal to 0.9 hours, and the MV-22B marked 1.4 hours, experiencing 552 failures in the 751.6 flight hours. Examiners found this statistic met not only the current threshold but that of the mature aircraft (one with 60,000 flight hours). Both parameters represented significant improvements versus the OT-IIE period.

As to maintainability, the two significant factors included the ratio of direct man hours per flight hour (MMH/FH) and mean repair time for aborts (MRT(a)). These ratios indicate how hard the maintenance people will have to work to support mission readiness and how quickly an aircraft with serious failures can be repaired and returned to flight status. The Osprey posted a remarkable 7.2 MMH/FH against a threshold of 20 hours or less (5,430 man-hours of direct maintenance), and an MRT(a) of 2.0 hours against a threshold of less than 4.8 hours.

Four other safety items were noted. The OPEVAL report found that the congested cabin and cumbersome seat belts might increase disembarkation time for the Marines and pose a safety risk during combat or emergency situations. It further

View from the cockpit as an MV-22 takes on fuel from a VX-20 HC-130. Refuelling can only be performed in the full airplane mode, and is generally conducted at 200 kt (370 km/h). With an engine out refuelling would be conducted at around 160 kt (296 km/h). Minimum airspeed in the airplane mode is around 120 kt (222 km/h).

found a restricted field of vision to the left from the cabin, which limited the ability of the crew chief to keep a safe lookout. The environment system was also found to be deficient as it did not adequately cool the cabin in hot weather. Finally, the V-22 cannot auto-rotate to a safe landing in the event of a dual-engine failure below 1,600 ft (488 m).

Regarding the entanglement hazards, the operational Osprey will deploy with a new crash-worthy troop seat design that eliminates the snag hazards. Regarding the environment system, an improved environmental control system was already planned for a Block upgrade. Regarding autorotation, it has been noted that there are other rotary-wing aircraft in the inventory that have little or no survivable autorotation envelope, particularly when heavily loaded. And the V-22 is less vulnerable to a dual-engine failure or combat loss due to the 45 ft (13.72 m) of separation between the engines.

Survivability

The report deemed the V-22 a survivable aircraft in low- and medium-threat environments, which it defined as assault rifles and larger calibre weapons (0.50-in/12.7-mm and 23-mm) adapted for anti-aircraft fire, and legacy MANPADS (SA-7 and variants). Much of the survivability portion drew upon testing conducted during the OT-IIE period. Indeed, susceptibility testing during OT-IIG focused specifically on RF and laser threats in order to examine corrected deficiencies from earlier tests. VMX-22 flew 15 sorties at NAWC China Lake Electronic Combat Range, and evaluated the radar and laser detection systems in terms of spatial coverage, detection range, threat identification, warning time and display accuracy. Tests were conducted against a variety of short-, medium-, and long-ranged surface-to-air missiles and acquisition radars.

The report concluded that the Osprey's integrated defensive electronic countermeasures system provided adequate capability for detecting radio frequency and laser energy directed at the aircraft. However, major deficiencies were noted in spatial coverage and threat displays that warranted addressing. The sufficiency of quantities and placement of countermeasure expendables for long missions was also mentioned. Although already planned for introduction in the Block B, the report recommended installation of a defensive gun.

CV-22B for AFSOC

The CV-22 programme had been on its own course since the mid-1990s. As the MV-22B flight testing progressed, planning continued on the basic CV-22 design and EMD effort. Interestingly, no EMD aircraft had been built to test the CV-22

Once the Osprey has arrived at its objective the task is to disembark the troops as quickly as possible. Normally the aircraft would land and the troops disembark by the back ramp (below). Another insertion technique is fast-roping (above), again conducted from the rear ramp. At present the Osprey is cleared for a 50-ft (15-m) hover, but that figure will decrease with further clearance work.

configuration. Thus, MV-22s Nos 7 and 9 were remanufactured to the CV standard beginning in the summer of 1999. Aircraft 7 was designated for TF/TA (terrain-following/terrain-avoidance) testing and to accomplish these tests, auxiliary wing tanks and radar were installed. Aircraft 9 was fitted with the Suite of Integrated RF Countermeasures, the TF/TA radar, plus the AVSS.

Following the operational pause, aircraft 7 resumed flight testing at Edwards AFB on 11 September 2002. Aircraft 9 underwent further modifications, incorporating additional MV-22 upgrades and did not return to flight until 14 July 2003. Additional modifications followed in the summer of 2004 to final production configuration. A third MV-22B (BuNo. 165839) was converted to a test asset in 2005.

The Air Force took delivery of the first production CV-22B (No. 20024) on 19 September 2005 at the Bell Helicopter production facility in Amarillo, Texas. After a short stint at Pax River for electromagnetic testing, the aircraft arrived at Edwards in December. The second production CV-22B (No. 20025) was delivered on 27 October 2005, and the third (No. 40026) was recently delivered in March 2006. The first two aircraft were production-representative test vehicles; the third was the first Block B/10, or in other words, the first CV-22 built to the Block B baseline configuration.

The Air Force Operational Test and Evaluation Center will use these three aircraft to conduct its Operational Utility Evaluation during the summer of 2006. If successful, the CV-22B will be certified for use in training operations at Kirtland AFB, New Mexico, beginning this autumn. The CV-22B initial operational test and evaluation is planned for October 2007, and initial operational capability is planned for 2009.

The V-22 today

The successful OPEVAL paved the way for the Osprey to enter full-rate production for both the Marine Corps MV-22 and the Air Force's CV-22. In its 29 September 2005 decision, the Defense Acquisitions Board (DAB) called for a gradual ramp-up of production with 2006 production remaining at the current 11 units per year, followed by 16 per year in FY07, 24 per year in FY08, and eventually reaching 48 aircraft per year

by FY2012. Negotiations for a five-year multi-year procurement spanning FY08-12 begins in August 2006. A total of 458 Osprey is authorised for production (360 MV-22s and 50 CV-22s), although the Navy's requirement for 48 MV-22s is still unfunded.

Bell-Boeing delivered the first Block B version of the Osprey to the Marine Corps on 8 December 2005 during ceremonies at Bell's Amarillo, Texas, facility. This Osprey (BuNo. 166491) marked not only the first MV-22B slated for a combat unit, but also marked the 19th aircraft delivered during 2005. As explained below, the Block B represents the combat capability that will deploy to fleet units.

Not only did the OPEVAL II document the Osprey's achieving all key performance parameters essential to its combat role, recommendations from that report validated the

An MV-22B from VMX-22 flies over the North Carolina coastal region. The engines are interconnected so that, if one fails, power can be shared by two propellers. In helicopter/conversion modes the engines are always at 100 per cent power, while in airplane mode they run at 84 per cent.

Named Chilly Willy, Aircraft 24 from the ITT was the aircraft that spent three winters running flying cold weather trials in Canada.

The tests at Nellis in mid-2005 were an important part of the OPEVAL. Here an MV-22 performs a landing in the desert (above), moving forwards slowly to avoid 'brown-out' conditions, while a ground team hook up a water container for an underslung lift – a typical battlefield task. For vertical operations the flaps are at full deflection: there are also 10°, 20° and 40° settings although the system is usually set to 'Auto', which automatically selects the optimum setting.

programme's roadmap for follow-on testing and evaluation to add further capabilities as the aircraft proceeds towards deployment.

In December 2005, Naval Air Systems Command conducted the first stage of an IOC Supportability Review to confirm the Osprey programme's ability to provide logistical support and maintenance for MV-22B operations. According to Marine Corps Colonel Bill Taylor, V-22 Osprey Programme Manager, PMA-275, "The OPEVAL establishes a baseline for us, validating all the work we've done to meet our obligations to the warfighters waiting on this capability." Taylor commented that the envisioned testing includes expansion of the Osprey's operating envelope for increased manoeuvrability, more extensive night and 'brown-out' operations, and tests to incorporate the M240 7.62-mm defensive gun at the rear ramp, a feature incorporated via the Block B improvements.

Early deliveries of the MV-22B Block B Osprey will go primarily to MCAS New River, where east coast Ospreys will be based. The anticipation is that a new tactical squadron will be transitioned about every six months, once the training has completed for the first squadron, VMM-263. VMM-162 and VMM-266 are scheduled to follow, leaving three tactical squadrons, the training squadron (VMMT-204) and the test squadron, VMX-22, operating by the end of 2008. Construction is underway at New River to accommodate the new aircraft, as well as a fourth Osprey simulator for the training squadron. All training will take place on MV-22B Block A configured aircraft.

VMMT-204 began receiving Block A aircraft from VMX-22 almost immediately after the full production decision in September, and in October the squadron launched its first training mission for a full class since the operational pause in 2000. Colonel Taylor commented that getting the training squadron back in the air is the first step in the training of a new generation of Osprey pilots and a "reflection of how far we have come with this programme in the last five years." Much of Colonel Taylor's remaining time as Programme Manager will focus on supporting the training squadron and preparing the operational squadrons that will take the Osprey into battle. Air Force CV-22B pilots will begin their training with VMMT-204, then move to the 71st Special Operations Squadron at Kirtland AFB, New Mexico, for training in specific CV-22 systems and tactics.

The CV-22B is scheduled for an Operational Utility Evaluation during the summer of 2006, to assess its capability to train pilots, thereby allowing Air Force pilots to train on the CV-22. An Initial Operational Training and Evaluation is then set for 2008 to assess specific mission equipment and capabilities introduced up to that point.

V-22 systems

The MV-22 Osprey is a multi-mission tiltrotor aircraft designed from the beginning for service as a common air vehicle for use among the four services. With the Army no longer looking at the Osprey, the two primary operators are the US Marine Corps and the Air Force, with hopes that the Navy will

eventually follow through on its unfunded requirement. The air vehicles for the two services are nearly identical, with the MV-22B Block B serving as the baseline for all future configurations. The CV-22B carries additional mission specific equipment for the special operations needs.

The Osprey combines the vertical and short take-off and landing features of a helicopter with conventional wing-borne flight of an airplane, resulting in a larger flight envelope than any previous assault support aircraft. This combination of air characteristics and the addition of aerial refuelling capability gives the Osprey the ability to self-deploy but also enhances survivability by allowing more flexibility in mission planning and threat avoidance.

The Osprey has three modes of operation, which are defined by the proprotor's incidence (the nacelle angle).

■ Vertical take-off and landing mode: flight with the nacelle angle above 85°

■ Conversion mode: flight with the nacelle angle between zero and 85°

■ Airplane mode: flight with the nacelle angle at 0°

The MV-22 was designed to be capable of taking off and landing with one operable engine, but cannot take-off or land in the airplane mode.

At present, the MV and CV configurations share roughly 90 percent commonality in airframe, 40 percent commonality in avionics and 100 percent commonality in propulsion. The differences are noted below and are mission-driven.

Common systems

Several core V-22B subsystems are common in both the MV-22B and the CV-22B. From an avionics standpoint, the Osprey has three fully redundant flight control systems, two fully redundant Advanced Mission Computers and a flight incident recorder. Communications are handled through two ARC-210(V) radios possessing voice and data comm., with SATCOM on both VHF and UHF channels with added anti-jam and encryption features. FM homing and UHF Automatic Direction Finding systems are provided as well. The Osprey also features a very high frequency (VHF) omni-directional range (VOR)/instrument landing system (ILS), a global positioning system (GPS) receiver, a Tactical Air Navigation (TACAN) system, and three redundant inertial navigation

systems, as well as a radar altimeter. Night operations are made possible through integration of night-vision goggles (NVGs), night-vision-compatible HUDS, and a forward-looking infrared camera display.

Electronic warfare is handled by the APR-39A(V) Radar Signal Detection Set, the AVR-2A Laser Detection Set (LDS) and the AAR-47 Missile Warning System (MWS). Sensors for the MWS and LDS are located on four quadrants of the

Above and left: VMMT-204 is the Osprey training unit, flying MV-22B Block A aircraft taken over from VMX-22. The unit received its first machines in September 2005 and the first course began the following month. In January 2006 the squadron loaned this aircraft to the first operational unit, VMM-263, to check the fit in the squadron's hangar. VMM-263 officially stood up in March and began receiving Ospreys in April.

Below: The third CV-22B for the US Air Force taxis at the test base at Edwards. The Air Force's Osprey differs primarily by having a nose-mounted blister for the APQ-186 terrain-following/avoidance radar.

Right: A CV-22B pumps flares from the rear sponson ALE-47 countermeasures bucket during a test flight from Edwards AFB. The aircraft has a number of non-standard test fxitures, including a forward-facing camera under the nose. Note the retracted refuelling probe. Self-protection is an important feature of the Special Forces CV-22, and it is fully equipped with radar, laser and missile warning systems. It also has the AAQ-24 DIRCM to counter IR-homing missiles.

Below: Two CV-22Bs set out from Edwards on an interoperability test mission. The Osprey is demonstrably quieter when approaching an objective than a conventional helicopter, greatly increasing its tactical surprise factor.

The first CV-22Bs (converted from MV-22 nos 7 and 9) were delivered to the USAF for trials in the standard light grey scheme (right, this aircraft also featuring the original fixed refuelling probe used to mount air data vanes). Later aircraft sported AFSOC's standard scheme of upper surfaces in a darker grey with a scalloped demarcation between it and the lighter undersides (above).

aircraft. Countermeasures are dispensed through the ALE-47 system. One dispenser is located in the aft portion of each sponson and dispenses a variety of rounds. Expendables may be fired manually or in one of six pre-programmed firing patterns. The CV-22B features an APQ-186 multi-mode radar, located in a pod affixed to the front port nose. The APQ-186 permits TF/TA flights in day, night and adverse weather. The radar features a ±40° scan in azimuth and +23/-40° in elevation. The CV-22B also has two additional ARC-210(V) radios and a Multi-mission Advanced Tactical Terminal (MATT). For added countermeasures, the CV-22B has two additional dispensable buckets in each of the aft sponson stations and two under the forward fuselage just aft of the nose gear. In place of the MV-22's APR-39, the CV-22 uses the ALQ-211 Suite of Integrated Radio Frequency Countermeasures (SIRFC), which automatically locates and classifies the various threats on the moving map display. The CV-22 also employs the AN/AAQ-24 Directed Infrared Countermeasures Systems to detect and defeat IR-guided threats.

MV-22B – Block A, B, C system

As discussed earlier, the improvements identified during the OE-IIE and operational pause were slated for incorporation into the MV-22B via a Block system, with Block A improvements meant to present a corrected training configuration for further testing and use in the final OPEVAL. The bulk of the Block A improvements centered around a redesign of the nacelle, with emphasis on safety and operational capabilities. Some of the noted capability improvements included improved lighting, modifications to the Osprey's cargo handling, inclusion of a fast rope bracket, and to the electronic hoist. The hoist is also mounted in a new location, over the ramp instead of the starboard door. Reliability and mainte-

nance improvements included modifications to the flat panel displays to make them more user-friendly. Substantial software upgrades were also included in the primary flight control system, control power management system and the mission planning system. Other cost-reduction measures were incorporated as well as training and flight manual improvements. Overall, the Block A focused on provided a safe and operational MV-22B.

Block B focused on enhanced maintainability improvements, and further incorporated the retractable refuelling probe, a production level anti-icing system, avionics/communications/navigation upgrades, and inclusion of the M240 ramp gun – one of the deficiencies identified in the OE-IIG report. Other efforts focused on improving overall maintenance by addressing parts obsolescence issues and additional cost-reduction changes.

The planned Block C improvements will see the addition of a flight incident recorder, weather radar, wheel well fire suppression and improved reliability cabin dome lights. Improvements are also in line for radar altimeter sling load interference and fuel dump capabilities, and a redesign of the main landing gear brake and tail position light. There will also be an electronic countermeasures system upgrade.

A Block D is currently under consideration, but is unfunded and unspecified.

CV-22B – Block 0, 10, 20 system

The CV-22B block upgrade is proceeding under a numerically designated system. It should be recalled that the CV-22B already incorporates the MV-22B Block A baseline capabilities. The CV-22B Block 0 is largely centred around inclusion of the multi-mode TF/TR radar and SIRFC. Other improvements see the inclusion of the ability to simultaneously talk and receive on all four ARC-210(V) radios, seating to accommodate the flight engineer, a multi-mission advanced tactical terminal, and interference cancelled. Five USAF expendable buckets will also be fitted to provide enhanced self-defence. Other capabilities enhancements include computer and digital map upgrades, a dedicated electronic warfare suite display, radar anti-icing features, an aircrew eye and respiratory protec-

tion system and the addition of 316 US gal (1196 litres) of internal and cabin auxiliary fuel tanks.

Block 0 has merged with Block 10 to form the baseline CV-22 aircraft. Additional avionics are designed to enhance the Osprey's ability to defend itself against infrared (IR) threats. Computer and software enhancements will improve the flight

Above: Osprey no. 7 is seen at Edwards shortly after completing the first terrain-following flight. The radar can be slaved automatically to the autopilot, or be used as an aid to hands-on flying.

Above and below: On 20 March 2006 the first CV-22B for Air Force Special Operations Command arrived at Kirtland Air Force Base, New Mexico. The aircraft will be flown by the 58th Special Operations Wing for special forces training after crew have been through the VMMT-204 syllabus.

Above: Having completed OPEVAL and passed some of its aircraft on, VMX-22 uses four MV-22Bs for continued operational evaluation and trials. Here two of the unit's aircraft operate from USS Wasp (LHD 1) in November 2005. The aircraft on deck is testing an unusual semi-metallic paint that has been trialled on a number of USMC types, including the AV-8B and AH-1.

Right: The tiltrotor is an obvious answer to the conflicting requirements of vertical versatility and straight-line speed, but the road to an operational Osprey has been a long and painful one. In 2000, after two tragic accidents, the programme came under real threat, but the Bell-Boeing team has recovered well to overcome the technical and funding difficulties to pass the OPEVAL test and receive a full-rate production decision.

Below: VMX-22 Ospreys taxi out at New River. The Block A aircraft will not be used operationally, being retained for training and test purposes.

engineer's display, navigation and digital map displays. Block 0 commonality with the MV-22B will drop to 88 percent airframe and 23 percent avionics.

V-22 missions

The Osprey gives the Marine Corps and the Air Force SOCOM a capability that currently does not exist. According to Colonel Taylor, putting the MV-22B capabilities in context demonstrates even more how the aircraft represents an improvement over the CH-46 that it replaces. Limiting the Osprey to the CH-46's radius of 75 nm (139 km), a 12-aircraft squadron of MV-22Bs can move a Marine battalion to location in one-fourth the time and in half the sorties. Moving a battalion in a CH-46 squadron would take 82 sorties and approximately 12 hours. Moving the same unit in an MV-22B equipped squadron would take only 41 sorties and approximately three hours.

When looking at the Osprey on its own terms, the result is simply not comparable with the CH-46. The Osprey brings an operational radius of more than 350 nm (648 km), which completely redefines the concept of force projection. Using the Osprey in this fashion would have significantly helped during the initial wave of operations during the US action in Afghanistan, when the insertion points were well beyond the range of the CH-46 and CH-53E. The aircraft had to be flown in a secondary mission role. As a further example, in Iraq the limited range of the CH-46 meant basing the squadrons closer to so-called 'hot' zones, thereby increasing their exposure to counter-attack. A squadron of MV-22Bs could have positions many miles to the rear, well out of harm's way, without jeopardising the troops' ability to quickly respond to contingencies.

Regarding the special operations mission, a US Navy SEAL team leader recently commented that in Afghanistan, the landing proximity to the assault target was defined by when the enemy would hear the approaching helicopters. Troops did not

fly directly to the target because by the time they would arrive, the enemy would have had several minutes preparation time to ready air defences. With the Osprey, these troops would have been able to proceed directly to the target, as the Osprey's speed and acoustic signature in airplane mode is so quiet that the enemy does not hear it until it is approaching the landing zone making its conversion.

The future

Although taking nearly 20 years to come to fruition, the V-22 Osprey has finally taken its place among operational US military aircraft. The Osprey programme has now moved from a developmental to a production programme and one of the most successful in the history of aircraft development. The Osprey is now poised to enter operational service, bringing

with it hitherto unavailable capabilities, which enhance force projection and increase safety for our troops. According to Secretary of the Navy, Dr. Donald C. Winter, "[The Osprey] proves that transformation is more than just a buzzword. ...

The combination of range, speed, and operational flexibility the Osprey provides is going to change all the rules on how our Marines engage the enemy."

Brad Elward

Below: Ospreys go to work – a sight that will become increasingly commonplace in the US forces.

F-103
retirement
"Adeus, Mirage"

Despite its replacement being some months away from service, the Brazilian Mirage III fleet was retired on the last day of 2005, leaving a temporary gap in air defence cover for the nation's capital.

The badge of 1° Grupo de Defesa Aérea featues the jaguar from the unit's nickname. For a while the group had two squadrons of Mirages but this was reduced to one in recent years due to dwindling aircraft numbers.

After 33 years of dedicated service, the Força Aérea Brasileira (FAB, Brazilian Air Force) retired its remaining Dassault F-103E/D Mirage IIIs on 31 December 2005. The aircraft – of which 32 were delivered between 1972 and 1997 – will be replaced by twelve surplus Dassault Mirage 2000Cs of the Armée de l'Air (French Air Force). The F-103 was of great national importance in South America's largest country as it was the first air defender of Brasilia, Brazil's newly founded capital. After the F-103 retirement, only two major air forces around the World will continue operation of this true classic of air defence fighters.

History

By the early 1950s the Brazilian government decided to relocate its government institutes from the popular beach city of Rio de Janeiro to the more centrally located and newly established city of Brasilia, which was inaugurated as Brazil's new capital in April 1960. By 1964, a coup removed the civilian government, and military rule was installed in the country. The generals concluded that the capital could not go without adequate military aerial protection and, following a competition between the English Electric Lightning, McDonnell Douglas A-4 Skyhawk, Lockheed F-104 Starfighter, Saab Draken and the Dassault-

Above: Ten F-103Ds were operated by the FAB, a mix of Mirage IIIDBRs supplied directly to Brazil and ex-French Mirage IIIBEs that were upgraded to IIIDBR-2 standard before delivery. Two of the original DBRs were also given the modifications.

Below: F-103E 4910 was the first Brazilian Mirage, and the first Mach 2 fighter in South America. It is maintained in immaculate condition at Anápolis, resplendent in the silver finish with Dassault's red house trim in which many Mirages were delivered.

Breguet Mirage IIIE, the latter was ordered.

On 12 May 1970, a $78 million order was placed for 16 supersonic Mirage IIIEBR/DBRs (BR – Brasil, in Portuguese), of which four were two-seaters – or 'bipostos' as the Brazilians call them. Just two days previous, the 1° ALADA (Ala de Defesa Aérea – Air Defence Wing) was founded near the city of Anápolis, located 150 km (93 miles) west-southwest of Brasilia. A completely new air base had to be constructed, receiving its first aircraft in August 1972. Meanwhile, in France, construction was completed of the first single-seat Brazilian Mirage (locally designated F-103E, two-seater became F-103D), FAB 4910. It first flew in March 1972. The FAB generals had sent eight pilots to the French Air Force Base Dijon for Mirage III training with EC 2/2. These pilots were – and still are – called the 'Dijon Boys'. These eight gentlemen were the first South American aviators to be supersonic jet operators, as Brazil was the first country in the region to order an aircraft with such capabilities.

The first F-103E, FAB 4910, was transported from France on a Brazilian Air Force C-130 of 1°/°1GT (Transport Group), together with another Mirage, to Base de Aérea Anápolis at the end of 1972. On 27 March 1973 Mirage 4910 took

The most obvious feature of the Mirage upgrade was the addition of canard foreplanes. However, small strakes were added either side of the tip of the radome on the single-seat F-103E. The radarless two-seater had larger strakes added either side of the nose.

to the air again in the hands of Dassault test pilot Pierre Varraut, signifying a notable milestone in the development of the FAB. Around the same time the GDA (Grupo de Defesa Aérea – Air Defence Squadron) 'Jaguares' was established in the 1° ALADA. Currently, FAB4910 stands proudly preserved opposite the Anápolis officers' mess. On 11 April 1979 the 1° ALADA was changed to the 1° GDA by presidential degree, part of a restructuring operation within the FAB.

To make good peacetime attrition and to augment the fleet the Brazilian Air Force received additional aircraft. Four more Mirage IIIEBRs were ordered in 1978 and delivered the following year so that a second squadron (2° Esquadrão) could be formed within 1° GDA. A spate of crashes in 1980/82, in which three trainers were lost, resulted in the hasty acquisition of two ex-AdA Mirage IIIBEs. Six further ex-AdA Mirages were ordered in 1987 – four Mirage IIIEs and two IIIBEs. Delivery began in October 1988 and the aircraft were modified to Mirage IIIEBR-2 and DBR-2 configuration, with canards. Two more IIIEBR-2s (4930 and 4931) and two IIIDBR-2s (4908 and 4909) were delivered in 1997.

Mirage operations

The Brazilian Air Force Mirages were primarily tasked with air defence, with a secondary air-to-ground role. Besides escorting Brazilian Presidents and foreign dignitaries, the 'Jaguares' were called on once to perform a real intercept. On 9 April 1982, two QRA F-103Es were scrambled to inter-

Above: For its secondary air-to-ground role the F-103 carried bombs on modified fuel tanks with four bomb racks each. Here the carrier aircraft is the 30th anniversary 'special'.

F-103D 4904 was given the name 'Ayrton Senna' after the legendary racing driver had flown supersonic in the Mirage. The aircraft was one of a handful of two-seaters left serviceable when the Mirage was retired in December 2005.

Above: A 1960s cockpit in the 21st century – this is the front seat of an F-103D. The 1990s upgrade changed the cockpit little, although an element of HOTAS control was installed.

Right and below right: The Mirages of 1° GDA were administered by IIIª Força Aérea (3rd Air Force), headquartered in Brasilia. Among the force's other assets are the A-1s (AMXs) and F-5s at Santa Cruz, and the F-5s at Canoas.

cept an unknown large aircraft that had entered Brazilian airspace. The Military Operations Centre (COPM) vectored the two fighters to a Cubana (Cuban National Airlines) Ilyushin Il-62. The Ilyushin was on its way from Havana to the Argentine capital, Buenos Aires, and carried the Cuban Ambassador to Argentina on board. It should be remembered that Argentina was in a state of war with the United Kingdom over the Falkland/Malvinas Islands at the time. Initially the Cuban crew ignored all directions ordered by the COPM, however, but they quickly changed their minds after seeing the 'hot' Mirages on their wings, which were serious about their intentions. The Cuban airliner was forced to land at the Brasilia International Airport, and was grounded accordingly.

Another proud moment in 1° GDA history occurred when the Brazilian Formula 1 racing driver Ayrton Senna was invited to fly with the 'Jaguares' on 29 April 1989. Mirage F-103D 4904 was used, under the command of Lt Col Alberto de Paiva Côrtes, then commander of the 1° GDA. Ayrton Senna Da Silva was tragically killed on 1 May 1994. Mirage 4904 carried his name, and still wore it in tribute when the aircraft was retired at the end of 2005.

1990s upgrade

When the 1988 batch of ex-AdA Mirages was delivered, the aircraft had been upgraded. This structural upgrade, part of the MirSIP (Mirage Safety Improvement Program) package, improved the aircraft's manoeuvrability by adding canard foreplanes and small strakes at the junction of the radome and pitot tube. These mods raised angle-of-attack limits from 28° to 40°. At the same time the aircraft were re-wired and single-point pres-

sure refuelling capability was added to dramatically decrease turn-round times. HOTAS controls were installed and the aircraft were made capable of firing the Matra R550 missile. Following the delivery of the four Mirage IIIEBR-2s and two IIIDBR-2s from France, the FAB then upgraded 10 existing F-103Es and two F-103Ds to a similar standard. The work was carried out in Brazil with Dassault assistance.

Other upgrades applied to the F-103s included the introduction of the Israeli-built Python III and the locally developed Mectron/CTA MAA-1 Piranha air-to-air missiles, plus structural upgrades for the newly assigned air-to-ground task. This fresh task had been assigned to 1° GDA in 1995 as a result of the introduction of new directives regarding the employment of FAB aircraft. To accommodate the new ordnance new drop tanks were developed that could each carry four Mk 82

bombs, like those developed earlier for the Mirage IIIOs of the Royal Australian Air Force. One F-103E (4929) was given a trial refuelling probe installation, but there were insufficient funds for a fleet-wide introduction.

The failed F-X

On 13 July 2000 the air force initiated a competition to replace the F-103, named the F-X. Aircraft competing for the F-X programme were the Lockheed Martin F-16C/D Block 52 Fighting Falcon, the Saab/BAE Systems JAS 39C Gripen, the Sukhoi Su-30/35 Flanker, the Mikoyan MiG-29 Fulcrum and the Dassault/EMBRAER Mirage 2000BR. While debate over the new fighter dragged on, in 2002 IAI proposed a 10-year lease of 12 upgraded Kfirs as an interim measure, but this was not taken up.

IAI had anticipated the complete cancellation of F-X, especially as it seemed that the Brazilian government and military were not entirely happy with their options. The US competitors were unable – due to US government rules on exports to South American countries – to deliver BVR-capable armament such as the AIM-120 AMRAAM. The Gripen and MiG-29 were perceived as being short on endurance. Dassault

The Mirage III not only holds a special place in the FAB's heart, but also in that of the local town, Anápolis. F-103E 4919 is on display in the town square.

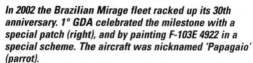
In 2002 the Brazilian Mirage fleet racked up its 30th anniversary. 1° GDA celebrated the milestone with a special patch (right), and by painting F-103E 4922 in a special scheme. The aircraft was nicknamed 'Papagaio' (parrot).

played it quite smart by acquiring a percentage of shares in the Brazilian aircraft constructor; EMBRAER, to market the Mirage 2000-5(BR). The favourite, with the FAB 'brass' as well as the pilots, was the Sukhoi option; however, the Brazilians lacked trust in their Russian counterparts.

As a result, F-X was indeed cancelled, and the Brazilians began in mid-2004 to look for an intermediate solution, with the long-term aim of acquiring a fifth-generation fighter after 2015. Following the change of mindset, the Brazilian Air Force looked at surplus MLU F-16s in the Netherlands, Mirage 2000Cs in France and IAI Kfirs. The latter had already been seriously investigated by FAB and 1° GDA crew who had already visited the IAI facilities at Ben Gurion near Tel Aviv, as well as the South African Air Force Base Louis Trichardt. It is known that FAB personnel flew the Kfir C10 in Israel as well as the Cheetah in South Africa.

After all the confusing cancellations and searches for a replacement for the aging F-103E/Ds of the 'Jaguar' Squadron, Brazilian President Lula went to Paris, accompanied by the

FAB display team 'Fumaça' (Smoke). On 14 July, the French national day, President Lula and the French President Chirac finally signed a Euro60 million contract that included the delivery of 10 basic RDI radar-equipped Mirage 2000Cs and two Mirage 2000Bs. The contract also included spares, overhaul, 12 new SNECMA M53-P2A engines and training for pilots and ground crew. Whether the acquisition of the Mirage 2000 will provide Brazil with full BVR capability is currently unclear. The aircraft will be delivered to a painstakingly slow schedule, with the first four deliveries expected by September 2006 and the last batch arriving in Brazil by August 2008.

Post-Mirage III era

On 14 December a major retirement ceremony for the F-103s in the Força Aérea Brasileira was held at BA de Anápolis. Squadron Commander Lt Col Carvalho Neto arranged a farewell flypast. Three of the original 'Dijon Boys' were carried in the back seats of the three remaining F-103Ds in the flypast. The last flight was accomplished on 31 December. The withdrawal of the F-103 has left a major air defence gap for the Brazilian capital. To fill it, the IIIª Força Aérea command has temporarily based a flight of four freshly upgraded Northrop F-5EM Tiger IIIs at Anápolis, drawn from the Canoas-based 1°/° 14 GAv (Grupo de Aviação) and the Santa Cruz-based 1° GAvCa (Grupo de Aviação de Caça). These Tigers are equipped with Python 4 and MAA-1 Piranha missiles.

The current group of Mirage III aviators will be divided from early spring as the first six 'Jaguares' will be sent to EC1/5 at Orange, France, to receive Mirage 2000 training. Currently all pilots are undergoing French language classes in order to follow their instructors while in France. Commander Carvalho Neto estimates that his first team will return by late August. On 4 September Brazil celebrates its independence day with a large parade in Brasilia. Every effort is being made so that the Mirage 2000 can take part.

The remaining 'Jaguares' pilots will be kept current on jet aircraft. For this purpose, some ageing EMBRAER EMB-326 Xavantes of 1°/5° GAv will deploy to Anápolis until all 1° GDA crew have been trained on their new Mach 2+ combat fighter.

Cees-Jan van der Ende and Iwan Bögels

The upgrade in the 1990s gave the F-103 fleet a new lease of life, but by the mid-2000s the aircraft was becoming increasingly expensive to maintain. The F-5 upgrade programme has provided aircraft with sufficient air defence capability to bridge the gap until the Mirage III's successor can enter service. Nine F-5s (six Es and three Fs) are being acquired from Saudi Arabia to augment the fleet.

509th
Bomb Wing
Whiteman Spirits

Since receiving its first B-2 Spirit at the end of 1993, the 509th Bomb Wing at Whiteman AFB, Missouri, has amassed many thousands of hours as the world's only stealth bomber unit. It has taken the B-2 to war over the Balkans, Afghanistan and Iraq, demonstrating the type's global reach, invulnerability to attack and precision attack capabilities. Here we review the capabilities of the B-2 and the organisation of the 509th Bomb Wing, and take a closer look at the Guam detachment, which began in 2005 as one of the 509th's important commitments.

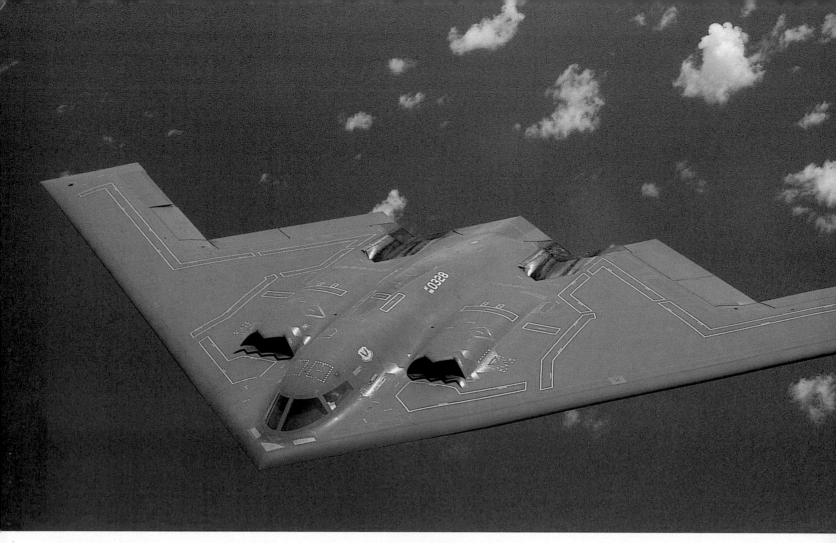

Commonly known as the 'Stealth Bomber,' the Northrop Grumman B-2 Spirit has now been around for nearly 18 years, having first flown on 17 July 1989. The maiden Spirit voyage, flown by serial number 82-1066, departed from Palmdale, California, at 0638 and landed approximately two hours later at Edwards AFB, California. Today, both Edwards AFB and Whiteman AFB, Missouri, have the unique distinction of being the only bases in the world hosting permanently based Spirits. Whiteman AFB received its first B-2, 89-0329 named the 'Spirit of Missouri', on 17 December 1993.

The unique flying wing is an extremely efficient design and, due to the lack of a typical fuselage and tailplanes, does not have the aerodynamic drag that other types of aircraft have. The unrefuelled range is approximately 9600 km (5,184 nm), and the low-observable technologies give the bat-winged bomber a freedom of action over hostile territory that other bombers cannot match. Low-observability enhances the aircraft's effective range as it is able to take a more direct route to the target, and it also allows for a better field of view for the aircraft's sensors. The B-2's low-observability is a result of reduced infrared, acoustic, electromagnetic, visual and radar signatures. The B-2 employs special coatings, composite materials and, when coupled with the flying-wing design, allows the aircraft's 'stealthiness' to be exceptional.

The B-2 has a crew of two pilots: an aircraft commander in the left seat and a mission commander in the right. The bomber sits 17 ft (5.18 m) high, is 69 ft (21.03 m) long, and has a wingspan of 172 ft (52.42 m). The gross take-off weight is 336,500 lb (152,636 kilograms) and the B-2 is able to carry a 40,000-lb (18144-kg) payload. The price tag for each B-2 was approximately $2.2 billion (price amortised over total programme cost of $44 billion) and that has been a highly controversial issue since it came into existence.

All 21 B-2s have been updated to Block 30 standards, giving the Spirit a fully functional weapons and navigation capability. The software continues to be updated on a regular basis and software/hardware upgrades are assigned a

A B-2A approaches the tanker over Utah during a training flight from Whiteman. The establishment of an overseas detachment at Andersen AFB provides the B-2 force with a new dimension for peacetime training.

Left: A Spirit slides away from the tanker after refuelling over the Pacific Ocean. Tanking is vital to the Global Power missions that are regularly practised by the B-2 squadrons, and which regularly reach over 24 hours duration.

Above: 'Spirit of Ohio', a B-2 previously assigned to the 412th TW but now with the 509th BW, flies past in slow configuration, providing good detail of the undercarriage and the 'drag rudder' surfaces that open above and below the wing tips for control.

P-number. There are a few P1.8 B-2s still, but most are at P3 (new communications suite) standard, and most recently the community just received some P4 upgrades. The pilots need to carefully watch what P-number aircraft of the three they are manning because there could be compatibility issues with pre-programmed devices before flight.

Initially Edwards AFB had the first six aircraft (82-1066 through 82-1071), but five were eventually transferred to the 509th BW. Edwards still retains number 82-1068 for testing purposes. Two test pilots and just the one B-2A are based at Edwards under the 412th Test Wing. The initial testing is done there and then modifications go to the Operational Test unit at Whiteman AFB. The B-2 radar modernisation programme testing was done at Edwards and was completed around February 2006.

While the B-2 was originally designed to have the potential to carry a third person behind the two front pilots, that capability was never pursued since it is usually filled up with equipment and also does not contain an ejection seat there. There is an area where a 6 x 4-ft cot can be placed for crew rest during long sorties and that is commonly done. Physiologists determine the crew swap-over times, sleep requirements and best times to eat, all built around the mission. Long non-stop missions could even last up to 45 hours if required. Standard local daily training sorties last 4-7 hours and each pilot must perform at least two Global Power sorties in a training cycle. Typical Global Power missions last around 24 hours.

B-2 weapons

The B-2 can carry both the version 1 and version 3 2,000-lb JDAM (Joint Directed Attack Munition), the version 3 being the penetrator model. The aircraft (P3 and P4 versions) can now carry 80 500-lb GBU-38 JDAMs, which is a newer weapon for the jet. The Spirit can also carry 16 2,000-lb Mk 84 GP bombs, the GBU-37 5,000-lb 'bunker-buster' that replaces the earlier BLU-113/GAM (GPS-Aided Munition), JSOWs (glide weapon), and most recently 16 JASSMs (advanced cruise missile). The B-2 ceased carrying cluster munitions along with the P3 upgrade, and the Spirit arsenal is now concentrating on the 'J-series' of weapons. While the B-2 could carry Mk 62 mines, that mission is unlikely to ever be

Many units claim to 'own the night', but few have as solid a stake as the 509th. Although its infrared- and radar-defeating properties are more crucial to its mission, the B-2 also has a low visual signature, especially at night and when viewed from head-on, behind or the sides.

Two views of 88-0330 at Andersen show the large bomb bay doors and airflow baffles that ensure clean weapons separation. A drop-down ladder provides access for the two-person crew.

carried out by a Spirit. Discussing nuclear weapon capabilities is somewhat taboo with bomber communities, however the B-2 is known to be able to carry both the B61 and B83.

The P4 upgrade now gives the B-2 a Link 16 capability, enabling the Spirit to share targeting information real-time with other assets while airborne. Another communication system upgrade is BLOS (Beyond Line Of Sight), which is linked to satellites and performs a similar task to Link 16. Link 16 was designed for line-of-sight operation, and is a virtual battlespace display in the cockpit: BLOS allows the B-2 to share information almost anywhere worldwide.

The B-2 is powered by four 17,300-lb (76.98-kN) thrust General Electric F118-GE-100 engines. A modification added a digital engine controller to maximise efficiency. The bomber operates within the high-subsonic speed category, and has a ceiling up to 50,000 ft (15240 m). Whiteman AFB has a hangar at the end of the airfield that is used for Spirit modifications. Northrop Grumman personnel occupy the hangar and a jet undergoing a mod such as a P-update is typically there between four to six weeks.

Whiteman AFB

One of the most memorable dates in the past 100 years is 7 December 1941. On that day, the Japanese attacked Pearl Harbor with a swarm of aircraft, crippling one of the most powerful navies in the world. When the enemy aircraft attacked, Second Lieutenant George A. Whiteman, a native of Sedalia, Missouri, scrambled to his Curtiss P-40 Warhawk at Wheeler Field, Hawaii. As soon as the P-40, nicknamed 'Lucky Me,' started to become airborne off the runway, he was downed by the enemy and was killed. His legacy still lives on today in the form of Whiteman AFB, Missouri, named as such in honour of the brave pilot.

The base began life as 'Sedalia Glider Base' on 6 August 1942. Waco CG-4A gliders, Curtiss C-46s and Douglas C-47s were based there. After the war, the base was deactivated and abandoned. In August, 1951, the base was reactivated as a Strategic Air Command base known as Sedalia AFB, with B-47s and KC-97s arriving for service in November 1953.

On 5 December 1955, the base was officially renamed Whiteman AFB. In June 1961, Whiteman was selected as the home of the fourth Minuteman wing, and received ICBMs in 1963. After 32 years of providing deterrence throughout the Cold War, the 351st Missile Wing was deactivated on 31 July 1995. Under START (Strategic Arms Reduction Treaty of 1991), the Minuteman missile system was eliminated.

On 30 September 1991, the 509th BW moved

Northrop Grumman B-2 Spirit fleet

Serial No.	'Spirit of'	Date rec'd at Whiteman	Notes
82-1066	America	23 December 1999	509th BW flagship*
82-1067	Arizona	18 November 1997	
82-1068	New York	n/a	based at Edwards AFB for testing, 412 TW
82-1069	Indiana	14 May 1999	
82-1070	Ohio	22 February 2000	
82-1071	Mississippi	7 April 1998	
88-0328	Texas	31 August 1994	
88-0329	Missouri	17 December 1993	13th BS flagship*
88-0330	California	17 Aug 1994	
88-0331	South Carolina	30 December 1995	
88-0332	Washington	28 October 1994	
89-0127	Kansas	17 February 1995	
89-0128	Nebraska	28 June 1995	
89-0129	Georgia	14 November 1995	
90-0040	Alaska	24 January 1996	
90-0041	Hawaii	11 January 1996	
92-0700	Florida	1 April 1996	
93-1085	Oklahoma	15 May 1996	
93-1086	Kitty Hawk	30 August 1996	
93-1087	Pennsylvania	4 August 1997	393rd BS flagship*
93-1088	Louisiana	7 November 1997	

*Flagships have no special markings other than the commander and DO's names on the nose gear shield. The wing flagship carries the commander's and vice-commander's names are on it.

from Pease AFB, New Hampshire, to Whiteman AFB in a non-operational status without people or aircraft. The 509th BW previously flew the FB-111A Aardvark for many years at Pease AFB. On 1 April 1993, the 509th BW became operational again. On 20 July 1993, the base resumed flying operations after a 30-year waiting period when the 509th BW's first aircraft arrived, a Northrop T-38A Talon, serial number 62-3690. The T-38s are used as companion trainers by B-2 pilots to keep their flying skills honed. The B-2s followed later that year, with the first example arriving in December 1993. As a result of base closures and realignment, Whiteman AFB also adopted the Air Force Reserve's 442nd Fighter Wing and its 22 A-10A Warthogs, following the closure of Richards-Gebaur Air Force Reserve Base, Kansas City, Missouri.

Whiteman AFB is located 65 miles southeast of Kansas City, and is two miles south of Knob Noster near US Highway 50. The base property includes 4,684 acres (1896 hectares) and has a 12,400 x 200-ft (3779 x 61-m) runway. The base is also responsible for a vast network of unused Minuteman II missile sites and launch control centres spread out over 3,300 sq miles (8550 km²) in central Missouri. Whiteman AFB is also home to a unit of Army AH-64 Apaches.

Units and aircraft

The B-2As at Whiteman AFB fall under Air Combat Command's 509th Bomb Wing. The aircraft are assigned to either the 13th Bomb Squadron, known as the 'Grim Reapers', or the

393rd Bomb Squadron, known as 'Tigers'. The 325th Weapons Squadron 'Cavemen' performs weapons officer training, the 72nd Test and Evaluation Squadron is responsible for OT&E, and the 394th Combat Training Squadron is responsible for T-38 and B-2 initial qualifications and requalification training. All five units share the 20 B-2As, resulting in a high demand for aircraft. In September 2005, the 325th BS 'Cavemen' was

Each pilot of the B-2 has four large screens that display a wealth of information. On long-endurance Global Power missions the crew take turns to nap during transit legs, ensuring both are at their best for important events such as refuelling and the target run.

redesignated as the 13th BS 'Grim Reapers', formerly a Dyess AFB, Texas-based B-1B Lancer unit. The squadron change was made to preserve

This KC-135R proudly displays its unit on the flying boom as a B-2 edges slowly forwards. The boom receptacle is located quite a way back on the B-2's spine. Note the frangible panels that form the canopy above the two pilots' seats, through which the ejection seats fire.

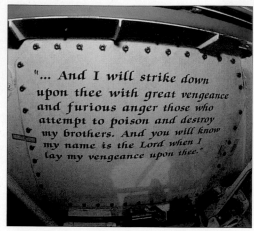

Carried inside access hatchway of 88-0328 is the biblical quotation made famous by Samuel L. Jackson's character Jules in the movie Pulp Fiction.

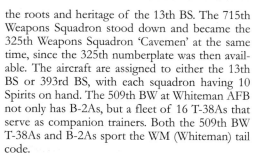

Above: A 412th Test Wing B-2 kaes a fly-by with its bomb bay doors open. The side-by-side bays can accommodate up to 80 bombs in the 500-lb (227-kg) class, and are long enough to admit weapons such as the GBU-37 'bunker-buster'.

Right: 'Tailcodes' and aircraft names are carried on the mainwheel doors. 'Spirit of New York' is the aircraft permanently assigned to the Edwards-based 412th Test Wing.

the roots and heritage of the 13th BS. The 715th Weapons Squadron stood down and became the 325th Weapons Squadron 'Cavemen' at the same time, since the 325th numberplate was then available. The aircraft are assigned to either the 13th BS or 393rd BS, with each squadron having 10 Spirits on hand. The 509th BW at Whiteman AFB not only has B-2As, but a fleet of 16 T-38As that serve as companion trainers. Both the 509th BW T-38As and B-2As sport the WM (Whiteman) tail code.

509th BW history

The 509th BW was activated on 17 December 1944 at Wendover AAF, Utah, under the command of Colonel Paul W. Tibbets, Jr. Its only mission in World War II was to drop the atomic bomb, which it did twice in 1945, ending the war. From July 1948 through December 1970, the 509th BW flew different types of aircraft that included the KB-29M, B-50, KC-97, B-47, B-52, and KC-135.

Some notable 509th missions during this period

included dropping an atomic bomb at Bikini Atoll on a captured and obsolete anchored fleet. The 509th also flew B-52 Arc Light combat missions over Southeast Asia in 1968 and 1969. December 1970 brought the FB-111A to the 509th, which the wing operated until 1988.

As previously mentioned, the 509th BW moved from Pease AFB to Whiteman AFB on 30 September 1991. The wing was again operational on 1 April 1993, but the individuals assigned to the 509th (known as '509ers') were actually working on the B-2 programme two years prior. The 509th BW Commanding Officer currently is Brigadier General (S) Greg Biscone. On 20 July 1993, the 509th took another important step when it received its first fixed-wing aircraft in almost three years: a T-38 complete with a B-2-style paint job. Later, the Wing's attention turned to the arrival of the first B-2.

Then, on 17 December 1993, the first operational bomber, named the 'Spirit of Missouri,' touched down on the Whiteman runway. Not only did the date mark the 90th anniversary of the first powered flight by the Wright Brothers, it also fell on the 49th anniversary of the original activation of the 509th Composite Group. With 20 B-2s now at Whiteman, the 509th continues to pioneer the operation of this unique aircraft.

As befits the mission for which the unit was established, the 509th Bomb Wing features a winged mushroom cloud as its badge. The unit's motto translates as 'Defender Avenger'.

The newest B-2 unit is the 13th BS, renumbered from the 325th BS. Its Grim Reaper emblem dates back to World War I when the unit (13th Aero Squadron) flew SPAD XIIIs in France with the US 1st Army.

The 393rd BS was the original squadron established to fly the B-29 atomic missions in the 509th Composite Group (the parent unit was not raised to wing status until November 1947).

Above: For an aircraft of its size the B-2 is surprisingly manoeuvrable, a fact underlined by the provision of a stick in the cockpit rather than a yoke.

Right: The 509th Bomb Wing's 394th Combat Training Squadron operates a fleet of Northrop T-38s, painted in a similar fashion to the B-2s. The Talons are important for maintaining pilot hours without racking up very expensive hours in the B-2 itself.

Some significant firsts associated with the B-2 include:

■ The first operational delivery of munitions by the 'Spirit of California' on 23 September 1994.

■ First B-2 appearance at a Red Flag exercise on 20 January 1995.

■ First B-2 flight to Europe by the 'Spirit of Missouri' on 10-11 June 1995.

■ The first B-2 mission over the Pacific by the 'Spirit of Kansas' on 1 September 1995.

■ The longest B-2 flight to date by the 'Spirit of Washington,' a 25-hour, non-stop, round-trip flight to Santiago, Chile, on 10 March 1996.

■ Three B-2s successfully executed the first live drops of the GAM on 8 October 1996 at the Nellis range complex. The bombers scored 16 kills with 16 munitions.

■ The first operational combat mission was flown on the first night of Operation Allied Force, 24 March 1999.

■ During Operation Allied Force the B-2s flew less than one percent of the combat sorties but dropped 11 percent of the total bombs.

Throughout the wing's history, its people, ever conscious of their proud history, realise their 509th ancestors established tough standards to follow. Still, '509ers' have every intention of equalling, if not surpassing, the past accomplishments of the 509th Bomb Wing.

Stealth Det at Guam

Andersen AFB, Guam, is an important staging base for the US military, and bombers are included. Guam, a US territory, provides the Air Force with an excellent location for strategic aviation assets. Various bombers based in the continental US are now rotated through Andersen AFB. While at Guam, units report to PACOM (Pacific Command) and to PACAF (Pacific Air Force), which directs a continuous bomber presence in the Asia-Pacific region. The bombers are a component of a deterrent that is designed to

Until the recent renumbering, the second B-2 squadron was the 325th BS, whose badge depicts a caveman riding a sabre-toothed tiger. The squadron number is now assigned to the Weapons Squadron.

This is the badge of the 36th Air Expeditionary Wing that parents the bomber deployments to Guam. Throughout the Cold War period the 36th (as a TFW) was the premier USAFE fighter wing in West Germany.

Aircrew assigned to the Edwards test fleet wear the patch of the 412th Test Wing. The single permanent test B-2 is allocated to the 419th Flight Test Squadron that is also responsible for B-1B and B-52H testing.

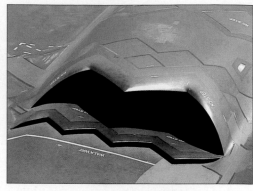

Above: *This close-up provides detail of the B-2's unusual intake arrangement. The lower sawtooth edge is the splitter plate that separates sluggish boundary layer air.*

Left: *During their stint on Guam the B-2s are tasked by PACAF as part of the theatre forces.*

promote peace and add security within the region. The bomber rotation in Guam has been going on since February 2004.

Bombers based in Guam give PACOM greater force projection as well as a global strike capability that previously did not exist. The continuous bomber presence at Andersen AFB was mandated by the Secretary of Defense and is not related to any one particular nation or event. Andersen AFB has always had strong community support and is strategically important, as it was during World War II and the Vietnam era. Flying in Guam gives units valuable deployment training and allows the personnel to participate in joint coalition training exercises.

When in Guam, bombers report to the 36th AEW (Air Expeditionary Wing) and the 36th Expeditionary Operations Group. Bombers are attached to Expeditionary Bomb Squadrons and use the same numeric squadron designation they use at their home bases. Bombers that maintain a presence in Guam include B-1B Lancers, B-2A Spirits and B-52H Stratofortresses. For rotations, units usually have a couple days of overlap built in to ensure there is 100 percent bomber availability

at any given moment. While the B-2s are much newer to deployments at Guam, they have recently received a 'FOL (Forward Operating Location) approved' status. Guam was selected as a FOL that the B-2s will now routinely deploy to. Two other FOLs include Diego Garcia in the British Indian Ocean Territory and Fairford, UK. The Guam detachment covers the Pacific Rim, Diego Garcia is close to Southwest Asia/Middle East, and Fairford encompasses Europe.

2005 detachment

In the summer of 2005 B-2As in Guam were deployed with their unit, the 13th BS 'Grim Reapers'. The 13th deployed its entire squadron, including 24 combat-ready pilots and more than 200 personnel. In April 2005, the 13th BS replaced the 393rd BS 'Tigers' who had since returned to Whiteman. With no less than four Spirits on hand at Andersen, there had been a B-2 presence on Guam for more than six months, making that deployment the longest on record for the B-2. During the deployment, the B-2s carried weapons such as 2,000-lb JDAMs and GBU-38 500-lb JDAMs for some of the missions.

In addition to flying standard training missions in Guam, the Spirits have also been flying Global Power missions that require accuracy and precision for 'time on target' scenarios that usually involves simulated targets that are far away. Throughout the Guam deployment the 13th EBS performed joint operations with the Navy during large force exercises in the area. B-2s operated alongside E-2Cs, F/A-18s and EA-6B Prowlers. The 13th also worked alongside Mountain Home, Idaho-based 391st FS F-15E Strike Eagles, and participated in other exercises including Kodiak Strike in Alaska. About 30 percent of the sorties involved integration training that included other aviation assets, and more than 25 percent of the sorties flown out of Andersen involved dropping weapons.

A B-2 pilot that was present for the deployment was Lieutenant Colonel Paul Tibbets. He is a fourth-generation namesake in his family and the grandson of Brigadier General Paul Tibbets, the pilot of the B-29 'Enola Gay' that dropped the first atomic bomb. At the time, that B-29 was assigned to the 509th Composite Group (later designated as a Bomb Wing) and the unit was commanded by Brig. Gen. Tibbets. Lt Col Tibbets has over 3,200 hours in various aircraft, with 800 hours in the B-2 including combat time in Operation Allied Force. Other aircraft he has flown are the B-1B Lancer and T-38A Talon.

Above: *Although flown by a 13th BS crew, 'Spirit of California' has the tiger-stripe and motif associated with the 393rd Bomb Squadron.*

Left: *Andersen deployments provide good opportunities to train with other units, especially those already based in the Pacific region and Navy squadrons. Here a 509th BW B-2 is escorted in a flypast over the base by a pair of F-15Es from the 391st FS at Mountain Home. The two disparate types trained together in various scenarios during their shared time at Andersen.*

Lt Col Tibbets said, "When you look at the capability that we have with the GBU-38, the 500-lb JDAM, it is incredible. My grandfather flew in enormous formations with thousands of men's lives at stake, perhaps trying to take out a single bridge or other strategic target. Today we can have a B-2 with two guys and eighty 500-lb JDAMs enabling us to surgically hit many more targets. Collateral damage is a minimal concern now and what we can do today is amazing. Mission planning is a lot more involved due to being able to select so many targets, and the target data can even be e-mailed to us after we are airborne using AMT (Airborne Mission Transfer). Our community is the only one that can do this at the moment.

"Low-observable airframes require more maintenance than traditional aluminum-manufactured aircraft. Working side-by-side with the maintainers during this deployment has been great since it is a closer-knit and tighter working environment – it

88-0330 gets smartly airborne from Andersen's runway, once the scene of intense B-52 activity during the Vietnam war. From Guam the B-2s can provide support rapidly throughout the Pacific region, notably to potential hot-spots such as Korea.

provides an excellent opportunity to get to know each other. We enhanced our working relationship with them and have learned to appreciate what they are doing for us. Even briefing with the tanker crews here, such as the Air Guard units that rotate through, gives us face-to-face opportunities to meet and work with others to get a better understanding of their mission. We have flown a record number of sorties from Guam and the maintainers have demonstrated great skill, attitudes and professionalism. That is important, since at Whiteman we report to two different groups that make us more isolated from the maintainers. Our flying schedule is somewhat designed around the maintenance schedule. The B-2s usually fly out of Andersen between three to four times a week and on seven-hour sorties, plus we fly between two to three Global Power sorties as well that last about 20 hours long for each one. The pilots fly two to three B-2 sorties a month.

"In Guam, we are the only game in town vying for B-2s to fly (at home we have four other B-2 squadrons all needing B-2 time), and it enabled us to remain focused and get excellent training which ultimately improves our capabilities. We became

A B-2 slips off the tanker over New Mexico. The accent of Spirit operations today is on long-range multi-target bombing sorties using 'J-series' weapons. These can be released simultaneously, but are independently targeted using GPS co-ordinates. For instance, if attacking an airfield a B-2 can release up to 80 JDAMs that can take out all the key points on the field such as tower, fuel supplies, intersections, hangars/shelters, defences etc – in a single pass.

involved in the local community and we supported two major events for the Guam Chapter of Habitat for Humanity. One event we provided labour for their annual 5K run fundraiser. The other event we helped them with was to finish building a house and then attended the dedication ceremony. It is important to give something back to the community and the base which have given us outstanding support."

Reduced mission times

Spirit pilot Major Brian Gallo has 2,400 hours total time with 600 hours in the B-2. He also has a lot of time in B-52Hs and T-38As. Major Gallo added, "One advantage of deploying to Guam is our mission may only take 12-18 hours to get to a

Left: Striding in after a mission are B-2 pilots Lieutenant Colonel Paul Tibbets (left) and Captain Ryan Bailey. Tibbets is the grandson of the 509th's first commander, and the captain of the first aircraft to drop a live atomic weapon.

Right: The B-2 will spearhead the USAF's strike/attack capability until a new vehicle is fielded some time in the 2020s. Discussions are ongoing as to whether that vehicle should be manned or not.

target versus missions in the 30- to 40-hour range that are flown out of Whiteman AFB. Flying for long periods is a difficult task and takes a pilot three days – one day to prepare, one day to fly it, and one day for recovery. When we fly Global Power missions, the training is very realistic. We take off, have to tank three to four times flying over the middle of the ocean, and we release weapons at a range very far away. It closely simulates an actual combat mission since you have to get the gas and drop weapons on an unfamiliar target.

"Being able to forward-deploy here and train in a region where we may have to fight someday is important. We learn, grow as a unit, and fight together – crews, support personnel and maintainers. It helps us know how to work together efficiently and makes our mission happen effectively.

A Pacific sunset provides a dramatic backdrop as a B-2 taxis for a nocturnal mission from Andersen. Deployments to the base foster enhanced co-ordination between aircrew and support staff.

You cannot do that well unless you deploy, practice and train. We are not using simulators or flying T-38s here, our missions are longer, and that gives us the opportunity to concentrate on the newer upgrade jets and become more fluent with our communications suite, including voice satellite, Have Quick II and HPW. Also we had an opportunity to practice missions with the 'new to the B-2' GBU-38 JDAMs and use the PRC-117 performing AMT, eliminating the need to perform mission planning before we take off."

Squadron commander

The 13th Squadron Commander is Lieutenant Colonel Tom Bussiere. Lt Col Bussiere flew F-15Cs and has now been in the B-2 programme since 1996. He has more than 3,000 hours total with 700 hours in B-2s, and has flown combat sorties in Operations Allied Force and Iraqi Freedom. When asked about Guam and the big picture, he said, "This is the first time B-2s have done an AEF deployment. Normally we are an

'AEF in place', which means we are on tap to deploy and we do the fighting from home base at Whiteman AFB. Last year the B-2 achieved FOC (Full Operating Capability) and part of that includes the ability to forward-deploy. This deployment is a culmination of the last 10 years by building up for our IOC (Initial Operational Capability) and Congress mandating FOC – with part of that requirement being forward-deploying. While we have deployed many times before, this is the first time we have deployed as part of a standard rotation. Going from ACC control over to PACAF control is fairly transparent and we simply become part of the theatre's assets when here.

"In the peacetime environment, Guam provides an excellent place to do training. The advantage of being deployed here during a conflict means we are much closer to the fight, allowing us to respond faster to any Western Pacific threats or tasking. The B-2 has been very reliable and some of the challenges during the Guam detachment have been the weather that was included typhoons and clouds of volcanic ash with gases that were emitted from a neighbouring island. The Air National Guard has been instrumental in providing tanker support enabling us to achieve our long-distance missions. The operators and maintainers have had a good opportunity to work closely together here and the longer duration training missions are a big bonus. The individuals in this squadron and everyone that has been deployed here have been great. While the airplane is amazing, it's our people that truly make it happen!"

Ted Carlson

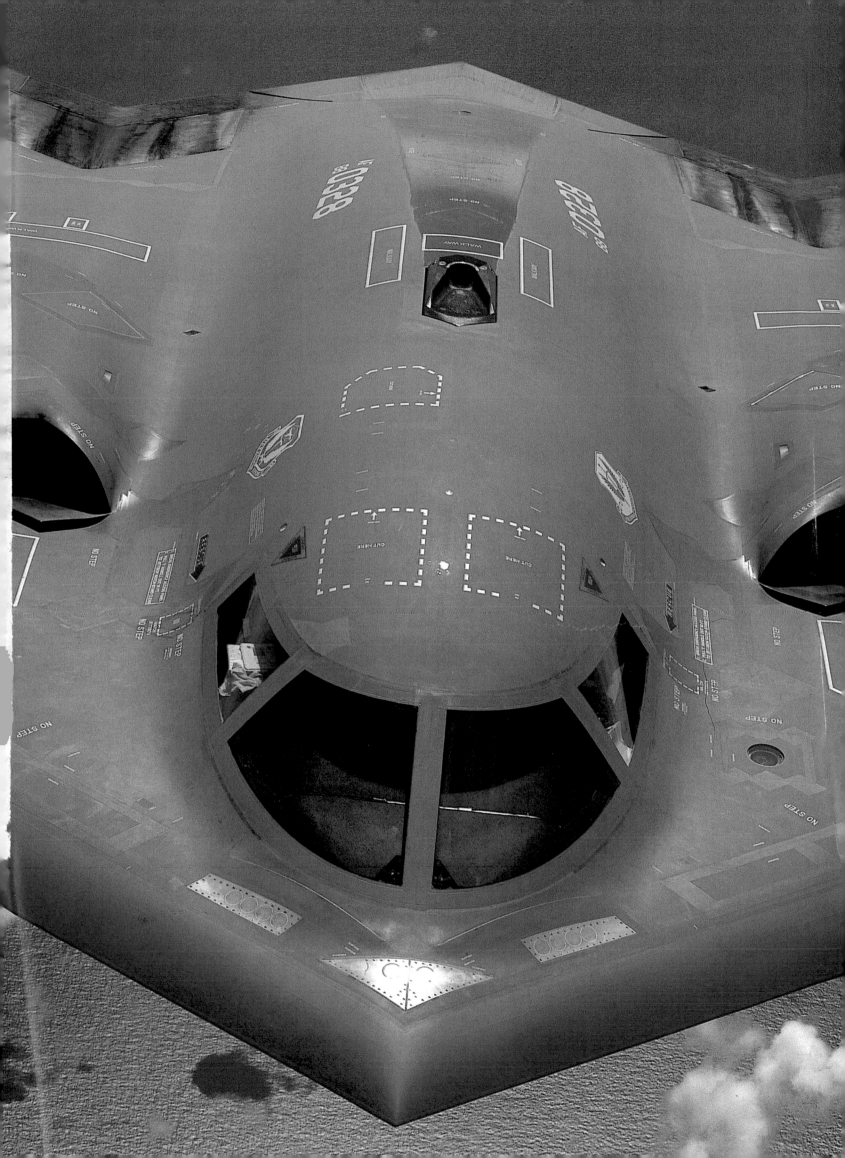

425 Squadron
Canadian Forces

Les Alouettes de Bagotville

Canada's armed forces are undergoing a transformation that moves them far away from their long-held Cold War posture. Leading the way is the CF-18 Hornet force, which is slimming down to two front-line 'super-squadrons'. Covering the eastern side of Canada is No. 425 Escadron Tactique de Chasse.

Currently Canada's Air Force is facing one of its most important challenges in its post-World War II existence. For decades Canada and its Air Force was focussed – as an ardent NATO-member – on defending western Europe, by stationing well-equipped fighter squadrons in West Germany and sending addi-tional combat squadrons to northern Europe in events of crisis or war. After the end of the Cold War and the inevitable cuts in military spending in the Western world during the 1990s, the Canadian Armed Forces reached a critical point, saddled with outdated infrastructure, an outflow of personnel, and a fleet of ageing aircraft.

The world's – and Canada's – views on national safety and international security were changed dramatically on 11 September 2001. Canada had to quickly and firmly respond to demands from the United States of America to bolster and enhance NORAD capabilities to defend North America against possible (terrorist) threats. Created in 1957, the North America Aerospace Defense Command evolved from a simple but vital response to the Soviet nuclear bomber threat to a more dynamic partnership handling multiple threats, ranging from airline hijacking, ballistic missile attack warning to space satellite monitoring. With the bi-national NORAD agreement due for renewal in May 2006, Canada and its Armed Forces needed to re-adjust their policies to avoid jeopardising vital military cross-border co-opera-tion by expanding NORAD's air defence-orien-tated profile to maritime surveillance and ground-based ballistic missile defence operations. Even a short-duration unilateral border closure by the US in reaction to or anticipation of a possible 9/11-type event would have a tremendous negative impact on Canada's export-orientated economy.

Step-by-step planning and sound forward thinking enabled Canada's Air Force to develop and gradually implement an innovative all-aspect transformation. Based on this official Air Force vision, "the Air Force will be transformed into an expeditionary, network-enabled, capability-based and results-focused aerospace force that will effec-tively contribute to security at home and abroad". At the same time, the Canadian Government

Bagotville – irreverently called 'Bagtown' – is now home to just one squadron of CF-18s. However, the merging of the two former units has created a single unit much better able to meet its commitments.

Left: Although it is downsizing in the process, the Canadian Hornet fleet is undergoing a timely upgrade process that will expand its capabilities with new radar and eventually datalinks and 'J-series' weapons.

Above: This Hornet carries a practice bomb dispenser under the port wing. The false canopy – to confuse opponents in close-in fights – has been a feature of the CF-18 fleet from the start.

decided for the first time in years to raise defence spending within the Federal Budget, enabling recruitment of additional personnel and some much-needed hardware replacements and updates.

With the defence of Canada and its inhabitants as primary concern, a new joint-forces military command, Canada Command or CanadaCom, was created in Ottawa to handle domestic operations, including search and rescue, disaster relief and, more importantly, anti-terrorism. To boost Homeland Defence capabilities, new aerial and seaborne surveillance hardware (UAVs and CP-140A Aurora update) was rushed into operational service. Simultaneously, a new CF Aerospace Warfare Centre (CFAWC) was established at Trenton, Ontario. This centre, initially staffed by experienced personnel drawn from various operational front-line units, will focus on doctrine development and will guide, streamline and evaluate the air force's ongoing transformation. The influx of experienced personnel from the various front-line units forced the Department of Defence (DND) to merge several CF-18, CP-140 Aurora and CC-130 Hercules units.

Finally, this new ambitious Defence Policy emphasises deployability (within Canada and 'out of area') and responsiveness as key elements. The Canadian Forces need to be able to get personnel and hardware where and when they are needed to protect Canada's interests.

One of the first front-line units to start the implementation of the Air Force's new vision and self-imposed mission statements is No. 425 (Alouette) ETAC (Escadron Tactique de Chasse, or Tactical Fighter Squadron), based at CFB Bagotville (Quebec) and flying the recently updated CF-18 Hornet. This Hornet 'super-

squadron', merged in July 2005 with the co-located No. 433 ETAC, is not only responsible within NORAD for the air defence of eastern Canada (and United States), but is also assigned to Canada's Expeditionary Forces, planning possible future deployment to Afghanistan.

Faced with this vast but vital operational workload, the internal transformation of No. 425 Squadron may well be a test case for the implementation and overall success of CF's Air Force Vision.

425 Squadron: CFB Bagotville

Since the receipt of its first CF-18 Hornet fighter-bombers in April 1985, No. 425 Squadron (which previously flew CF-101 Voodoo interceptors) was re-roled from a pure NORAD-assigned air defence/interception unit to a multi-role

fighter-bomber squadron based at CFB Bagotville in Quebec. Located in a remote part of this French-speaking province, No. 425 Squadron joined No. 433 (Porcupine) Squadron as part of No. 3 Wing. The latter squadron, receiving its CF-18s in December 1987 as a replacement for its obsolete CF-116 (CF-5) Freedom Fighters, was assigned to NATO and would have deployed to Norway in event of tension in western Europe.

Having become fully operational on the new CF-18 Hornets, both independent squadrons staffed the Bagotville-based NORAD 'quick reaction area' and occasionally intercepted Soviet Tu-95 'Bear' long-range patrollers heading for

Two CF-18s take off from Bagotville. No. 3 Wing no longer maintains a permanent air defence presence at Goose Bay, although No. 425 Squadron still practises deployments to the Labrador base.

A CF-18 pilot climbs into his mount. The Canadian Air Force faced something of a pilot shortage in the 1990s, but the reduced size of today's force and the merging of units means that most squadrons are now fully staffed.

Cuba. The withdrawal and deactivation of the Germany-based CF-18 squadrons after the closure of CFB Baden-Sollingen in 1993 enabled both squadrons to integrate experienced 'expat' CF-18 pilots in their ranks, some of them having gained precious operational experience during the Desert Storm war over Kuwait and Iraq. In 1999 Bagotville pilots also participated in NATO's Operational Allied Force (known as Operation Friction to the CF) over Serbia and Kosovo, operating out of Aviano AB, located north of Venice.

Once returned to Bagotville, both squadrons quickly resumed their day-to-day flying operations. From time to time, both squadrons would spend time at their Deploying Operating Base at CFB Goose Bay, Labrador, and their Forward Operating Location at Iqaluit in the Northwest Territories province in northern Canada. The arctic FOLs (Inuvik, Rankin Inlet, Yellowknife and Iqaluit) are well-equipped with operational infrastructure (drive-through aircraft hangars) to house CF fighter squadrons during exercises or crisis situations 'up north'. The western FOLs (Inuvik, Rankin Inlet and Yellowknife) are traditionally assigned to No. 4 Wing at CFB Cold Lake, Alberta, home base of the CF's two remaining operational CF-18 units (Nos 416 and 441 Squadrons). Furthermore, CFB Cold Lake houses

also CF's dedicated CF-18 conversion unit (No. 410) and the AETE test and evaluation unit.

The implosion of the Soviet Union and its armed forces, and the vanishing threat of Soviet bombers frequently harassing CF and NORAD air defence forces, influenced No. 3 Wing's operational workload during the 1990s. At the same time both squadrons needed to retain their operational capabilities with ever-decreasing financial defence budgets.

NORAD alert

Within hours of the terrorist attacks on 11 September 2001, NORAD aerospace control contingency plans were put into effect. As a direct response six fully-armed CF-18 Hornets of No. 425 Squadron were dispatched to CFB Greenwood, Nova Scotia, to patrol the east coast aerial approaches to North America. Additional Bagotville CF-18s were put on alert at their home base and at CFB Goose Bay, Labrador. The CFB Cold Lake squadrons were sent to western Canada (especially to CFB Comox near Vancouver) to monitor the western aerial approaches to North America. Since September 2001, several CF-18s are still launched daily to patrol areas of Canada. Although the nature and exact profile of these missions (including maximum reaction time) remain highly classified, the concentration of the country's vital economical and political installations and urban areas bordering the St Lawrence river and the Great Lakes may well justify these daily patrol flights. Until recently, the 'QRA-birds'

were also launched during weekends. During 2005, the CFB Goose Bay Hornet QRA, staffed by Bagotville pilots, was deactivated, pending immediate re-activation if necessary.

Implementation of the Air Force vision

As already mentioned, Canada's Air Force embarked in mid-2005 on an ambitious transformation focused on an improved ability to react to suddenly emerging threats, both within its national borders and abroad. Since surveillance, air defence and deployability are key factors to achieve this transformation, the Canadian Minister of Defence decided to merge several squadrons to achieve organisational and operational benefits of scale.

Two CP-140A Aurora maritime patrol squadrons (Nos 415 and 405) at CFB Greenwood, and two CC-130E/H transport squadrons (Nos 429 and 436) were forced to merge. At CFB Bagotville No. 433 Squadron, being activated 13 months later in 1943 than its Bagotville sister unit, was consolidated into No. 425 Squadron on 15 July. A similar – but not enforcing – request to merge two CF-18 squadrons was sent to No. 4 Wing at CFB Cold Lake. Willing to closely monitor the squadron merger at CFB Bagotville prior to starting its own in-house reorganisation, No. 4 Wing elected to postpone the consolidation of its two operational CF-18 squadrons.

Fortunately, the Bagotville merger did not trigger a decrease in number of aircraft, pilots and flight time allocated to No. 3 Wing. Faced with two mostly 'undermanned' squadrons, the decision to merge all 'airborne' assets not only allowed the Wing to re-group all available personnel but also to achieve increased productivity and aircraft availability.

Staffed by 24 operational CF-18 pilots, originating from both previous squadrons, the 'new' No. 425 Alouette Squadron is responsible for the air defence of Canada. Secondly, the unit needs to be capable of sending at short notice two deployable 'six-packs' (i.e. six CF-18s and 12 pilots) to an out-of-area location to safeguard Canada's interests abroad. Although a far cry from the two deployable CF-18 wings, envisaged by the Government's 1994 White Paper, the improved operational capabilities of the recently updated CF-18s, capable of using 'smart' bombs, should compensate for the decreasing number of available aircraft. With no formal governmental and political decision taken until now, the DND has already tested various deployment scenarios to send additional military 'assets' to Afghanistan in support of CF's Task Force Afghanistan. If sent

Canada's Hornet simulators are an important part of the training and operational evaluation processes. The facilities at Bagotville and Cold Lake are linked so that pilots can fly missions simultaneously, allowing the practice and development of advanced tactics and multi-ship attacks.

Flares are released during a training mission. In the Cold War the roles of the various Hornet units were more clearly defined, but today the units train for true multi-role missions in a variety of scenarios.

to Afghanistan, the Canadian CF-18 Hornets and their aircrew will most likely be based at Kandahar airfield in the south of the country.

In line with Canada's Air Force vision statements, the 'merged' and fully-staffed No. 425 Squadron decided to focus itself from the start on its operational core businesses: air defence and deployability. All non-operational second-line operations and missions were 'out-sourced' to special-purpose units, created within No. 3 Wing. The operational training of new CF-18 pilots, type-rated by the CFB Cold Lake-based No. 410 Squadron, is organised by the new CEAC/Centre d'Entrainement Apte au Combat (Combat Ready Training Unit), staffed by four to six instructors when needed. In the course of four to six months, students receive an advanced CF-18 combat-ready training syllabus (including air-to-ground and air-to-air tactics, NORAD procedures and inflight refuelling) before being integrated in No. 425 Squadron.

As a multi-role fighter-bomber unit, the squadron flies on average 60 per cent of its missions for air-to-ground training over Quebec. The introduction of the new updated CF-18 Hornets enabled the squadron to relinquish its low-level attack mission and to shift to medium-level attacks, using the aircraft's capabilities to drop 'smart' bombs. Nowadays, low-level awareness training is mostly flown in the flight simulator at Bagotville. A familiarisation flight at low altitude is flown in a dual aircraft with an experienced CEAC instructor or squadron pilot in the back-seat. Air-to-ground gunnery missions are flown to the nearby Valcartier Army gunnery range, or similar ranges near Petawawa, Ontario, or Gagetown, New Brunswick.

In early September 2005, No. 425 Squadron hosted a bi-national large-force employment exercise at CFB Bagotville. During four days, No. 425's CF-18s trained in various air defence scenarios

Wearing 425 Squadron's alouette (lark) badge on the fin, a two-seat CF-18B taxis at Bagotville. The two-seaters are fully combat-capable, but are also used for training. At Bagotville the CEAC instructs students fresh from type conversion at Cold Lake in the intricacies of combat flying. This focuses the main squadron's activities on operational flying and training.

with six US F-16C Fighting Falcons (Vermont ANG), four F-15C Eagles (Massachusetts ANG), two A-10 Thunderbolt IIs (Connecticut ANG), three KC-135Es (Maine ANG) and a NATO E-3B AWACS aircraft. Several missions included 12 fighter aircraft countering simulated attacks by six 'enemy' aircraft. Every year the entire squadron deploys to CFB Cold Lake to participate in the well known and realistic multinational Maple Flag exercises, attended by a large number of European air forces, as well as Canadian and US squadrons.

Winter 'migration'

To fine tune its air-to-air training, No. 425 Squadron annually deploys to the United States to train with various USAF, US Navy and/or Air Force Reserve units. This is usually conducted in the winter to escape the harsh Quebec weather conditions. In February 2006 the unit migrated to Eglin AFB, Florida, to fly Alouette Mobile DACT (Dissimilar Air Combat Training) missions with the F-15C Eagles of the 33rd FW and F-16C Fighting Falcons of the co-located 85th Test Squadron/53rd Wing. Afterwards, No. 425 participated in a large-scale air defence exercise at Homestead AFB, Florida, homebase of the 93rd FS 'Makos' of the 482nd Fighter Wing (AFRC). During this Chumex 2006 exercise, the CF-18 aircrew trained with the local Air Force Reserve F-16C pilots, joined by USMC AV-8B Harriers from MCAS Cherry Point and F-15C Eagles from Tyndall AFB. A similar Combat Archer deploy-

Several aircraft retain mission markings from Operation Allied Force/Friction. Eighteen CF-18s were deployed to Aviano for the campaign, flying 678 sorties in both air-to-air and air-to-ground roles, the latter using a mix of dumb bombs and Paveway II laser-guided bombs.

ment to Tyndall in 2005 enabled the Bagotville pilots to shoot AIM-7F Sparrows and AIM-9P Sidewinders at unmanned BQM-34A drones over the Gulf of Mexico.

Finally, before its possible deployment to Kandahar, the squadron will participate in NATO's multinational Brilliant Arrow 2006 exercise, operating out of Skrydstrup air base in southern Denmark. Since the CF's own strategic Airbus CC-150 Polaris tankers of CFB Trenton (Ontario)-based No. 437 Squadron are still undergoing modification, No. 425 Squadron frequently

CF-18 Hornet weapons

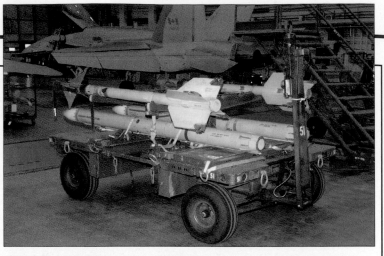

Above: The AIM-9 Sidewinder and AIM-7 Sparrow remain the standard air-to-air missiles for non-upgraded Hornets. Modified aircraft can carry the AMRAAM.

Below: These are MUTACTS pods, which allow air-to-air fights to be replayed in detail for accurate post-mission debriefing.

Above: The M61A1 Vulcan 20-mm cannon is still a very useful weapon for both ground strafing and air-to-air work. In the CF-18 it is provided with 570 rounds.

Below: A common store is the combined practice rocket/bomb dispenser, although it is normally only used for carrying bombs. The small bombs mimic the ballistics of full-size weapons, and have a small smoke charge for visual scoring.

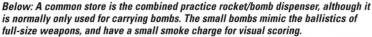

trains with and uses the KC-135E tankers of the 101st Air Refueling Wing 'Maineiacs'/Maine ANG, based at 'nearby' Bangor IAP for air-to-air refuelling during training and ferry flights.

Similar to other CF units, No. 425 Squadron includes a number of foreign exchange pilots. For several months, a USAF F-16C pilot and the RAF's Flt Lt Dan 'Dingle' Ingall, an experienced No. 6 Squadron Jaguar pilot and qualified weapons instructor, have worked as operational pilots in the unit. Operating within a mutual 'full access – no disclosure' agreement with the Royal Air Force (or other air forces), Dan Ingall (and all non-American exchange pilots) is not allowed to become familiar with prevailing NORAD procedures and cannot pilot QRA Hornets under NORAD control. By contrast, US and Canadian exchange pilots are allowed to man European QRA fighter aircraft while on exchange with vari-

ous European NATO air forces. One of the No. 425 Squadron pilots started his military pilot career as an operational F-16AM Fighting Falcon pilot with the Dutch Koninklijke Luchtmacht. Sent to CFB Bagotville as a Dutch exchange pilot in 1999, he selected to stay in Canada at the end of his exchange tour, coinciding with the end of the air force contract with the Koninklijke Luchtmacht. Having received a similar contract with the Canadian Armed Forces, he nowadays is one of most experienced pilots in the unit.

For air-to-air training, all Hornet aircraft are equipped with US/Swiss-designed MUTACTS (Memory Unit Tactical Air Crew Training System) air combat instrumentation pods, enabling detailed and visual debriefing by all participants minutes after landing. This system, also in use by the US Navy, Finnish and Royal Malaysian Air Forces, provides accurate reconstruction of

complex CF-18 Hornet training sorties without the need for an instrumented range or any additional equipment on the aircraft

On occasion, No. 425 Squadron also trains with Canadian naval forces and CP-140 maritime patrol aircraft. Operating out of CFB Greenwood, east coast homebase of the CF's Aurora MPA fleet, CF-18s fly escort missions with the Auroras during joint exercises with Navy destroyers and/or frigates. One of the more elusive naval support missions is flown by the squadron after a retrofit of the Navy's surface vessels. To certify the retrofit, CF-18s of No. 425 zoom at supersonic speed over the Navy vessels for the 'obligatory' post-retrofit acceptance shakedown.

CF-18 Hornet upgrade

To successfully perform this multitude of missions within an Air Force in full transformation, No. 425 Squadron nowadays operates a mix of 'old' and recently updated CF-18s. Out of 138 CF-18 Hornet aircraft initially purchased in the early 1980s, 121 airframes remained in service. Only 92 Hornets will be kept in operational service until the termination of Phase 1 of the update programme, planned in 2006. This contract, awarded in 2001 to Boeing, involves the in-depth multi-phased update of 80 airframes to F-18C/D standard, based on the US Navy's Engineering Change Proposal (ECP) 583. The remaining redundant and unmodified aircraft will be used for spare parts or offered for sale to other air forces.

Upgraded Hornets like this aircraft make the force much more useful for multi-national operations. Phase II of the upgrade will significantly enhance their capabilities with 'J-series' weapons, Link 16 and a new targeting pod.

Although the home defence role has changed dramatically in nature since 2001, the CF-18 force still practises deployments to the northern FOLs. This was once a crucial element in NORAD's defences against the threat of Soviet 'over-the-Pole' attacks.

The first updated CF-18, modified by the Montreal-Mirabel based L-3 MAS (Military Aviation Services) company as a sub-contractor, was delivered to No. 441 Squadron at CFB Cold Lake in 2003. The first phase of this CF-18 modernisation included the installation of AN/APX-111 IFF, more capable AN/APG-73 radar, two Have Quick radios and embedded GPS/inertial navigation systems, and the capability to fire AMRAAM BVR (Beyond Visual Range) active-radar guided missiles. Officially still known as ECP-583 CF-18s, the Air Force Command preferred to start equipping all front-line units at the same time with updated Hornets. The first aircraft were delivered to Nos 416 and 441 Squadrons at CFB Cold Lake, soon to be followed by Nos 425 and 433, and finally to the conversion unit, No. 410. Since both 'variants' have identical aerodynamic and pilot-input characteristics, it was not necessary to send experienced CF-18 pilots to No. 410 for re-orientation on the new ECP-583 jets before the receipt of updated Hornets by the front-line squadrons.

With more updated CF-18 Hornets arriving at CFB Bagotville, the CEAC is nowadays exclusively instructing 'freshman' CF-18 pilots on the ECP-583 aircraft. On 6 January 2005, the 40th updated CF-18 Hornet was delivered to the Canadian Armed Forces, ahead of the financial initiation of Phase II of the programme on 18 February.

The second phase, to start in June 2006 and be completed in 2009, will see the integration of Link 16 secure communications and new colour cockpit displays. The Link 16 will enable the aircrew to stay in constant contact with other aircraft, ground stations and/or AWACS aircraft, and will hugely improve the combat situation awareness of the pilot. The introduction of a helmet-mounted optical sight system will enable the pilot to visually

While the Hornet force goes through the upgrade process squadrons are operating both new and old machines. The ECP-583 jets (left) can be distinguished from unmodified aircraft (right) by the addition of the antenna array forward of the windscreen.

monitor the target without having to look down for vital targeting information. During Phase II, the CF-18 cockpit will be modified to become fully NVG (Night Vision Goggles)-capable, greatly enhancing night flying and attack capabilities. In July 2005, test pilots of AETE deployed to British Columbia to test the use and assess the flight safety impact of the CF-18 NVGs.

New Hornet equipment

In addition to the ECP 583 update programme, the CF is studying the purchase of a more advanced thermal targeting pod to replace the ageing Nighthawk pods, and GPS-guided munitions (Joint Direct Attack Munition/JDAM and/or Joint Stand-off Weapon/JSOW). The CF is planning to buy up to 50 new targeting pods in 2006, under the Advanced Multi-Role InfraRed Sensor (AMIRS) project. Main contenders are the Litening-AT advanced targeting system, PANTERA (Precision Attack, Navigation and Targeting with Extended Range Acquisition) and the Advanced Targeting FLIR (ATFLIR) systems.

The CF is seen as a possible customer for the KEPD 350 long-range air-to-surface missiles, already selected by the Spanish Ejercito del Aire to arm its EF-18 Hornets and Typhoon fourth-generation fighters. The CF recently decided not to procure ASRAAM short-range infra-red air-to-air missiles due to their high price.

To optimise operational training on the new CF-18 Hornets, the Department of National Defence (DND) purchased six Bombardier CF-18 Advanced Distributed Combat Training Systems (ADCTS). The ADCTS implements a system of networked CF-18 cockpits at Cold Lake (Alberta) and Bagotville (Quebec). Being able to detect each other visually on the simulator screens and electronically by radar, team training between the various squadrons – despite being physically separated by a great distance – is made possible within a simulated but realistic operational environment. Furthermore, to improve the training value of these high-tech simulation systems, ADCTS can be linked to similar Navy and Army – and even US-based military – simulators for joint training, at a 'marginal' overall cost.

Once the Phase II ECP-583 CF-18 is in full operational service, from 2009 on, the CF will have (most likely) one fully operational CF-18 squadron at CFB Bagotville and another at Cold Lake, fulfilling its multitude of missions to defend Canada and the country's overseas interests.

Stefan Degraef and Edwin Borremans

Vickers Supermarine Swift

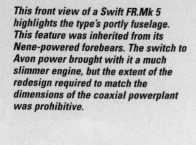

This front view of a Swift FR.Mk 5 highlights the type's portly fuselage. This feature was inherited from its Nene-powered forebears. The switch to Avon power brought with it a much slimmer engine, but the extent of the redesign required to match the dimensions of the coaxial powerplant was prohibitive.

Designed and built by Supermarine, whose wartime Spitfire had made the company the UK's premier fighter manufacturer, the Supermarine Swift was the RAF's first swept-wing jet fighter, the first RAF aircraft with power-operated ailerons and the first in regular service with reheat. The Swift also broke a number of speed records in its time, and even briefly held the world absolute speed record, reaching a speed of 737.7 mph (1187 km/h) at low level over the Libyan desert, though this record was broken only hours later by the Douglas Skyray. Later, the Swift enjoyed a brief but glorious career as a low-level tactical reconnaissance aircraft, a role in which it was said to have been 'unsurpassed', piling up the silverware in a number of NATO competitions, and winning an enviable reputation for operational efficiency. Little of this is remembered today, and the Swift is now most often recalled as a failure, frequently finding itself in lists of the 'World's worst aircraft.'

The Swift was ordered into production as a safeguard against the possible failure of the Hawker Hunter, the RAF's preferred jet day fighter at the time, at the height of the Cold War, and at a time of unequalled tension between the Communist Bloc and the Western allies. The aircraft were urgently required to replace Fighter Command's outdated and obsolescent Meteors and Vampires. Both aircraft suffered severe problems when first introduced into service, but while the Hunter's were quickly solved, allowing it to become one of the most successful post-war fighters, the Swift's problems proved so serious that the fighter version had to be withdrawn in May 1955, only 15 months after its introduction, and before it could enter full operational service.

Orders for 492 Swift fighters were cancelled, and just 15 aircraft reached the single front-line unit to receive the type – a squadron which never completed conversion to the ill-starred Swift.

However, the Swift's record was better than its reputation might suggest. Though the fighter was abandoned before it could become operational, more than 100 FR.Mk 5s were delivered, and these equipped two front-line reconnaissance squadrons, gaining a reputation for effectiveness in the role. Although 20 Swifts were lost in flying accidents, only four of these major accidents were fatal. In these accidents, the front-line squadrons lost just 13 aircraft (three F.Mk 1s and 10 FR.Mk 5s) and only two pilots. This was a remarkably good record by the standards of the day, and eight of the FR.Mk 5s lost were abandoned after engine failures – hardly the fault of the much-maligned Swift.

Evolution

The Swift evolved from the straight-wing Attacker (via the Supermarine Types 510, 517, 528 and 535), gaining a 40° swept wing, a modern nosewheel undercarriage and, at the last moment, an axial-flow turbojet engine. The latter should

Supermarine's first jet fighter was the Attacker – a first-generation, straight-wing jet that drew heavily on the Spiteful/Seafang (a development of the Spitfire) and packed a bulky Nene engine into its barrel-like fuselage. The first stepping stone to the Swift was this prototype, the Type 510, which flew with swept wing and tail surfaces.

have allowed a move away from the Attacker's broad fuselage contours, but in the event time was too pressing for this to be achieved, and all production Swifts had a bigger, fatter fuselage than was strictly necessary.

Many of the personnel who created the Swift were members of the team that had previously designed and built the immortal Spitfire. Joe Smith, chief draughtsman under Mitchell, and Chief Designer after Mitchell's death in 1937, had been the father of the Spitfire, and would go on to design the Swift. There was also a real and direct link between the Spitfire and the Swift, via the often forgotten Supermarine Spiteful. The Spiteful had been directly developed from the Spitfire, by adding a laminar-flow wing to a late-series Spitfire fuselage. Though the Spiteful and its naval cousin, the Seafang, were rendered obsolete by the dawn of the jet age, their laminar-flow wings were used to form the basis of Supermarine's first jet fighter, the E.10/44.

This was originally due to enter RAF service as an interim type pending the introduction of the Gloster E.1/44, but the aircraft offered little improvement over the Meteor and Vampire, and the RAF version was cancelled. The aircraft did fly in prototype form, however, and its naval version was the Attacker, which did enter service, becoming the Royal Navy's first front-line jet fighter. The first E.10/44 made its maiden flight on 27 July 1946, and the type entered Royal Navy service on 17 August 1951.

Though its tailwheel undercarriage and unswept, plank-like wing appeared archaic, the Attacker flew rather better than it looked and, though it was not fast enough to contest the World Speed Record, the aircraft did achieve a number of point-to-point records. On 27 February 1948 Mike Lithgow flew the Attacker prototype to set a new speed record of 570 mph (917 km/h) around a 100-km closed-circuit course.

Swept-wing research

To meet RAF requirements for a Meteor replacement, the Air Ministry issued specification F.43/46 for a new jet-powered daylight interceptor. It was clear that any such aircraft would be fitted with swept wings, and to explore swept-wing flying characteristics experimental specifications were issued to Hawker and Supermarine,

Above: The Swift was seen as an interim swept-wing fighter to be fielded ahead of the more promising Hunter. In the event, the failure of the Swift, delays to the Hunter and the pressing need for fighters that could counter Soviet MiGs led to the hasty procurement by the RAF of large numbers of Canadair-built Sabres.

Below: A rare colour shot depicts the Supermarine Type 510 VV106 flying along the English south coast. Given that it was a trials platform for a new and relatively unknown technology, the Type 510 performed well, exhibiting more than enough promise for continued development and further prototypes.

each covering the design and construction of two prototype aircraft. To expedite the process, these experimental research aeroplanes were to be derived from existing types – Hawker adapting the P.1040 (later Sea Hawk) to meet Specification E.38/46 and Supermarine adding swept surfaces to the Attacker under E.41/46. The Hawker machine was designated as the P.1052, while the Supermarine aircraft became the Type 510.

Supermarine's design team and experimental shop at Hursley Park near Romsey (to where it had been evacuated during the war) immediately set to work on the Type 510. This aircraft was basically

an Attacker with swept wings, retaining the Attacker's archaic tailwheel undercarriage, and powered by a centrifugal-flow Nene 2 engine. This dictated the Attacker's rather fat fuselage contours, which were wider than those of the rival Hawker P.1067. The first Type 510 (VV106) made its maiden flight at Boscombe Down on 29 December 1948 in the hands of Mike Lithgow, about a month after the rival P.1052. Boscombe Down's long runways were often used for first flights and special tests, not least by Supermarine, whose test airfield at Chilbolton was relatively small.

Above: With its naval Attacker origins, it was no surprise that the Type 510 was tested on board an aircraft-carrier. With strengthened undercarriage and an arrester hook fitted, VV106 landed on Illustrious on 8 November 1950, the first swept-wing jet to land at sea.

Below: The Type 510 was designed with a Sabre-like slightly tapered wing. As the problems associated with swept wings manifested themselves, the Swift's wings evolved so that by the time of the Mk 5 they had kinked leading and trailing edges, plus dogteeth.

With a maximum design speed of 700 mph (1126 km/h), the Type 510 was the first British jet aircraft with both swept wings and a swept tailplane. Though archaic, it marked a major step forward for Supermarine. The company had only delivered the last Spitfire 10 months before, on 20 February 1948.

Soon referred to as the Swift, the Type 510 demonstrated easy and pleasant handling characteristics, with a limiting Mach number of 0.93 imposed by a marked lateral trim change (requiring full opposite stick to counteract) above that speed. Turn performance at altitude was poor, but the aircraft had none of the Dutch roll and snaking then expected of swept-wing, high-speed aircraft. The Boscombe Down test pilots who flew the aircraft expressed misgivings at the tailwheel undercarriage, but reported that with a tricycle gear and improved elevators, the Swift had

"the makings of a good fighter." Supermarine did finally have an aircraft capable of beating the Mach 0.92 demonstrated by the Spitfire!

Better was to come when the RAE undertook a comparative evaluation of the Type 510 with the F-86A Sabre. At low level, the Type 510 achieved 520 kt (963 km/h), compared to the Sabre's 580 kt (1074 km/h), but at 25,000 ft (7620 m), the Swift's 617-mph (993-km/h) top speed was just 3 mph (4.8 km/h) slower than the Sabre. Moreover, because the Type 510 had been designed primarily to gain experience with swept wings, the aircraft's surface finish was poor, and Supermarine was confident that a production aircraft would be appreciably cleaner, with significant performance benefits. In the meantime, the two Type 510s underwent a succession of modifications and were redesignated as the Type 517, the Type 528 and the Type 535.

Real improvements were achieved with later variations on the Type 510, most notably the Type 535, which made its maiden flight on 23 August 1950. The Type 535 had a modern tricycle undercarriage and an afterburning Nene engine, the latter the product of a private venture programme by Supermarine.

Though the early 'on/off' two-position afterburner – with its vertical 'eyelid' nozzle – was crude, it did the job, and the Swift itself finally 'looked the part' enough to star in the major cinema film *Sound Barrier*, in which it played the part of the fictional 'Prometheus' fighter. Supermarine had finally caught up with Hawker, whose all-swept P.1081 flew first, but which crashed before reheat could be installed.

Though there were proposals to use the P.1052 and P.1081 as the basis of operational fighters, Hawker's chosen operational fighter design was the P.1067, which would eventually become the Hunter. This soon became the RAF's preferred aircraft, too, and in March 1950 the Ministry of Supply ordered an initial batch of production Hawker P.1067s, even before the first prototype had flown.

The Supermarine Type 535 was seen as a useful back-up plan for the Hunter programme, providing a potential fall-back in the event of any problems with the Hunter. After the Korean War broke out in June, it was believed that a production Swift could be in front-line service earlier than the Hunter, and it became inevitable that the type would be ordered into production to augment the Hawker aircraft. As the Hunter programme 'wobbled' there was even brief pressure to cancel the Hawker aircraft in favour of the Swift.

Many believe that a productionised Type 535, with a Nene engine and wing-mounted cannon, could have been a very fine interim fighter, marking a massive improvement over the Meteor, and available without undue risk. More time could then have been spent ironing out the problems on the definitive Avon-engined version. It was quickly decided, however, that any production Swift would be powered by the axial-flow Rolls-Royce Avon, and not by the Nene. It was felt, however, that the Swift would still be a low-risk programme, thanks to the progress already made with the Type 510/517/528/535.

'Super Priority'

Once the Korean war was underway, attention was focused on getting the aircraft into service in the fastest possible time. In November 1950 the aircraft was ordered into production for the RAF under the 'Super Priority' scheme, even before the Avon-engined Swift had flown in prototype form, and less than a month after the two Type 541 Swift prototypes were ordered. Specification F105P called for the production of two prototype E.41/46 aircraft, based on the Type 535 but with RA.7 Avon engines, and for the manufacture of 100 production fighters. Supermarine was given the initial contract to produce the first 100 aircraft, but another 100 were ordered from Short and Harland in 1952. Orders for 300 more from each factory were placed in January 1953. Offshore procurement was expected to take the total to more than 1,200 aircraft. Supermarine insiders later recalled that, when the first order for 100 Swifts was placed, no-one had a clear idea of what a production Swift would look like.

The Operational Requirement that followed (OR228) called for Mach 0.95 (547-kt/ 1013-km/h) performance at 45,000 ft (13716 m),

with a time to that height of less than six minutes and a service ceiling (at which a climb rate of 1,000 ft/305 m per minute was to be possible) of 50,000 ft (15240 m). This was ambitious stuff, and the Air Ministry willingly accepted a two-cannon armament scheme in an effort to speed up development. No serious thought was given to producing a Swift powered by the alternative Armstrong Siddeley Sapphire engine, as happened with the rival Hunter, however. Ominously, theoretical studies by the RAE predicted that the Swift would be entirely unable to meet these performance requirements, without a major redesign of the wing, though this conclusion seems to have been ignored.

Swift production

While the first Swifts were notionally built at Supermarine's South Marston plant, in reality components and sub-assemblies came from a baffling array of workshops, part of a system of dispersed production inherited from the wartime Spitfire programme. Noses were built by Thomas Harrington Ltd, "Motor Coach Builders and Automobile Engineers" of Hove, while Boulton Paul in Wolverhampton manufactured wings. Even within Supermarine, different factories played a part in building Swifts. Though assembled at South Marston, the Swifts' fuselages incorporated components from Eastleigh, and the wings had parts from Itchen.

From the start, it was expected that these fighter Swifts would be built to four different standards, the initial two-gun F.Mk 1s giving way to four-gun F.Mk 2s, and then to reheat-equipped F.Mk 3s, and, "at the earliest possible date" to the F.Mk 4 with an all-flying tail. The latter aircraft was expected to be released to service by January 1953.

Placing such large orders without completing development was a disastrous mistake, but no-one expected the Swift to suffer such severe problems, and with war raging in 'MiG Alley' over the Yalu River, and with Fighter Command still utterly dependent on Meteors and Vampires, time was of the essence. In December 1951, the RAF's Operational Requirements Branch was seriously expecting deliveries from October 1952.

Having flown briefly as the Type 528 afterburner-equipped tail-dragger, VV119 was modified to Type 535 standard with an extended nose housing a nosewheel (below). A curved fin fillet was later added (above) and in this form it starred as the 'Prometheus' fighter in the movie Sound Barrier.

WJ960, the first Type 541 prototype, flew on 1 August 1951, but engine and other problems (including two dead-stick landings) caused major delays. On one occasion, the engine failed on finals, and pilot Dave Morgan was unable to reach the airfield. Despite coming down outside the airfield, wheels-up, and despite hitting an apple tree and a brick-built privy on landing, the aircraft was soon back in the air. The second prototype, WJ965, which was far more representative of the production configuration, flew on 18 July 1952 but continued to experience problems. With spring tabs removed Dave Morgan dived the aircraft to Mach 1 on 26 February 1953, though before the trim tabs were removed, severe aileron flutter had rendered the aircraft almost uncontrollable above 550 kt (1018 km/h). On several occasions, flutter caused such a distortion in the fuselage that the undercarriage lowered, provoking a sudden nose-up pitching moment sufficient to cause the pilot to black out.

By then, production aircraft were starting to pour off the production lines, the first production Swift F.Mk 1, WK194, having made its maiden flight on 25 August 1952. This was less than six weeks after the second prototype had flown, and long before problems revealed by the flight test programme could be solved. This aircraft was formally accepted by the Ministry of Supply on 31 October 1952.

Customer concerns

The customer, in the person of the Aeroplane & Armament Experimental Establishment's Squadron Leader Chris Clark, finally flew WK194 on 8 November 1952. Clark's report, dated three days later, listed a litany of concerns, but A&AEE requests for a full preview trial were rejected by the Ministry of Supply on the grounds that Supermarine had remedial work in hand.

The type was then evaluated by a USAF team with a view to including the type in the offshore

Superficially the Type 541 Swift prototype looked similar to the Type 535, but there were important differences, not least of which was the switch to the Rolls-Royce Avon. The aircraft also had a revised wing of greater span.

procurement programme that, it was suggested, could see 'hundreds' of Swifts being built in Holland. By 1953, *Aviation Week* was reporting orders for some 375 Swifts. In fact, the USAF team felt that it could not recommend the Swift because of wing drop, directional instability and poor controllability at high Mach numbers.

Instead of allowing the customer access to the aircraft, Supermarine pressed on with its own priorities, and from 1 June no fewer than eight F.Mk 1s were deployed to Boscombe Down, with four A&AEE pilots, two from the RAE and two from the CFE, not for intensive trials, but to prepare for the flypast that was planned to accompany the Queen's Coronation review of the RAF at Odiham on 15 July 1953.

Coronation review flypast

The F.Mk 4 prototype, which had first flown in May 1953, and WJ965 were among the six Swifts that took part in the flypast. By mid-1953, the Swift was then suffering an embarrassing run of engine failures, when Hunters using the same engine were not. It was assumed that the fault lay in the Swift's engine intakes, and enormous efforts were made to find a solution.

The five Swifts in the flypast were reportedly fitted with new engines for the event, but WK198 suffered an engine seizure afterwards when approaching Chilbolton. Pilot Mike Lithgow managed to pull off a successful forced landing. The day after the flypast it transpired that the fault was not the Swift's – Supermarine had received Avon 105 engines whose turbines had come from a particular sub-contractor who had made an apparently trivial change to the compressor blade root fixings, and this had caused all of the failures. After a minor modification to the blade roots, the problem was solved.

The F.Mk 4 made its name by breaking the world absolute speed record over Castel Idris, Libya, on 26 September 1953 with Mike Lithgow at the controls. The aircraft made four runs over the course, following a series of smoke pots – achieving a speed of 735.4 mph (1183.5 km/h), and beating Neville Duke's record of 727 mph (1167 km/h) set using the one-off Hunter F.Mk 3 three weeks earlier. Two runs made during a subsequent attempt averaged 743 mph (1196 km/h), but the afterburner then failed, making the requisite four runs impossible. The record only lasted a day, though, and even before WK198 returned to Chilbolton the record had been taken by American's Douglas Skyray, which pushed it up to 742.7 mph (1195.2 km/h) on 27 September, and then to 752.94 mph (1211.7 km/h) on 3 October.

Before the type's most serious weaknesses were exposed by service use, the Swift's troubles seemed to be no more than a natural function of the rushed development programme, and the type seemed to offer greater potential than the Hunter. *RAF Flying Review* confidently quoted that: "the Supermarine fighter holds promise of being able to carry us very much further along the sonic path. And there is little doubt that RAF Fighter Command will be the first to agree that the Swift has been well worth waiting for." Orders kept piling up, so that the Select Committee on Estimates later said: "As the difficulties with the aircraft increased, so, it appeared, did the number on order."

With demands for the first Swifts to be delivered to the first squadron before the end of the year, efforts were made to gain a proper Controller Aircraft release, and the second Type 541, WJ965, was fully instrumented and sent to Boscombe Down. Within the month, the aircraft had been lost, when its pilot failed to either recover from a spin or eject. WK201 and WK202 were despatched to the A&AEE's 'A' Squadron in January 1954 but the directive was that the CA Release should be achieved as quickly as possible, and should be based on the bare essentials necessary for the aircraft to be used for 'crew familiarisation'.

The conventional procedure of completing manufacturers trials, then A&AEE trials and clearance, and then CFE/AFDS trials, before issue to a front-line squadron were put aside for the Swift in the rush to get the type into squadron service. There were still major problems with the Swift, which limited what could safely be undertaken.

A restricted release was issued on 12 February, limiting the aircraft to 500 kt (926 km/h) below 5,000 ft (1524 m), then to Mach 0.9 above that, and to a maximum height of 25,000 ft (7620 m). Spinning was prohibited. The aircraft was not judged able to fulfil its intended role safely, and was therefore limited to non-operational use. As if

The second Swift prototype was far more representative of production aircraft, and was used to speed along the release to service trials although it was lost at Boscombe Down in the process.

WK194 was the first Swift F.Mk 1, flying soon after the second prototype. Such an accelerated programme meant that the early Swifts were beset with numerous problems that might have been fixed if the programme had been conducted at a more realistic pace.

WK194 was the first Swift F.Mk 1, flying soon after the second prototype. Such an accelerated programme meant that the early Swifts were beset with numerous problems that might have been fixed if the programme had been conducted at a more realistic pace.

this were not bad enough, most pilots disliked the aircraft's basic handling, one A&AEE test pilot derisively referring to the aircraft as a 'Blowtorch-driven breeze block'.

The Central Fighter Establishment's Air Fighting Development Squadron looked at the Swift while No. 56 Squadron was converting to the new fighter, instead of before, and its conclusions were not inspiring. The AFDS professed itself "thoroughly disappointed" by the poor elevator control, and judged the surging and compressor stalls "totally unacceptable" in an operational aircraft, though they were impressed by the ability to scramble the aircraft quickly, thanks to its quick and easy start-up and excellent rate of climb. Visibility from the cockpit was poor, and the aircraft was felt unable to cope with Venoms or Sabres in simulated combat. It was judged that "extraordinary improvements" would have been necessary to turn the aircraft into an acceptable high-level interceptor. The AFDS report was finally issued on 28 June 1954, by which time No. 56 Squadron had been operating the type for four months.

Squadron service

The 'Firebirds' received their first Swift F.Mk 1s the day after the clearance document was signed, on 13 February 1954, and by the end of the month 'A' Flight had received nine aircraft. These were the first swept-wing jet fighters in RAF service, but although A&AEE continued trials work aimed at extending the release, No. 56 Squadron continued to operate under the original limited clearance, and were subject to numerous restrictions with regards to gun firing, g, top speed (550 kt; 1018 km/h or Mach 0.91) and service ceiling (not above 25,000 ft).

One aircraft was abandoned in a spin on 7 May, and another was lost in a fatal crash on 13 May, necessitating changes to the aileron control circuit. The aircraft were grounded for two months for modifications and 'A' Flight reverted to the Meteor F.Mk 8. Swift flying resumed in August, but another aircraft was abandoned on 25 August, after the nosewheel failed to extend. As a result of this incident, all Swifts were grounded in late August. The surviving F.Mk 1s were augmented by F.Mk 2s from 30 August, but this new variant proved even less useful. When the Type 510 had been evaluated against an F-86A, initial impres-

sions had been favourable, and the British aircraft's performance had been competitive, but the stability problems encountered by the F.Mk 2 rendered it virtually useless as a front-line fighter.

Unfortunately, when an extra pair of cannon were added, the space required for the extra ammunition was made by extending the wing leading-edge roots forward at the fuselage. This changed the shape of the wing, with a modest kink at roughly mid-span. This caused pitch-up problems at high Mach numbers, leading eventually to a high-speed stall, or to a rapid and violent departure if the controls were mishandled.

Some 45 trial modifications to the wing failed to solve the pitch-up problem, and instead extra ballast had to be added to the noses of the F.Mk 2s. This made the aircraft heavier and much more sluggish, and less easy to handle on landing, while the type retained all the shortcomings of the F.Mk 1. Nevertheless, the A&AEE allowed its use, in the same non-operational role, and subject to the same limitations as were defined in the F.Mk 1's release. Fifteen different Swifts eventually saw service with No. 56 Squadron, but there were never more than eleven on charge at any one time.

It was hoped that a second Fighter Command Swift squadron (perhaps the co-located No. 63 Squadron) would form with the Swift F.Mk 3s, but they were destined never to be flown by the RAF, being delivered straight to storage before being issued as ground instructional airframes.

Furthermore, testing had by then revealed that all was not well with the F.Mk 3. It turned out that the early Swift's reheat would not light up at high altitude, and the aircraft had to descend from its typical operating altitude of 35,000 ft (10668 m) to 20,000 ft (6096 m) in order to light the afterburner. For an interceptor, this was entirely unac-

ceptable, and the F.Mk 3s went straight into storage at No. 15 MU.

Meanwhile, Supermarine continued working on the further improved F.Mk 4 (Type 546). This had an all-moving, variable-incidence tailplane, which finally seemed to cure the aircraft's pitch-up problem, though the reheat still proved problematic and the wing still produced insufficient lift, so that realistic fighting capability was limited to altitudes below about 40,000 ft (12192 m), and so that the Swift was outclassed by other fighters above about 15,000 ft (4572 m).

Serviceability problems

Quite apart from the Swift's problems in the air, the aircraft was something of a nightmare for its long-suffering engineers and groundcrew, with appalling serviceability, and even by 1955 it was rare for No. 56 Squadron to have more than half of its aircraft serviceable and on the flightline. It was this, more than the aircraft's limitations, that led to its unpopularity on the squadron and with the RAF 'brass'. On 8 February, the Chief of the Air Staff informed the Secretary of State that it would require "an unduly heavy amount of maintenance" to keep the Swift "operationally serviceable" and advised against accepting any of the Swift fighter marks for service. He also ruled out the high-altitude PR.Mk 6, but left the door open for the proposed FR.Mk 5 and F.Mk 7.

On 23 February 1955, Prime Minister Harold Macmillan was forced to explain the Swift saga in the House of Commons. Macmillan admitted

Gleaming in the Wiltshire sun, Swift F.Mk 3s sit outside the South Marston works pending delivery. Such were the problems with the few earlier Swifts that had reached the front line that the entire Mk 3 batch was delivered into storage at nearby Wroughton.

WK198 was modified from a Mk 1 with the F.Mk 3's wing and a new variable-incidence tailplane to become the F.Mk 4 prototype. It also tested additional wing fences (left). With Mike Lithgow at the controls, the aircraft briefly held the world absolute speed record. Eight production F.Mk 4s were built, of which four were converted to FR.Mk 5 and one to an F.Mk 7.

that: "It has been decided that the Marks 1, 2 and 3 cannot be brought up to an acceptable operational standard, while after further tests on the Mark 4 it has been decided that production of this mark should be restricted to a limited number." On 15 March 1955 the order was given to cease all Swift flying forthwith, and preparations were made to ferry the remaining aircraft to Maintenance Units.

Two aircraft (WK207 and WK210) were ferried to No. 33 MU at Lyneham on 28 March, followed by WK206 the next day, while six more (WK221,

239, 240, 242, 244 and 245) followed on 31 March. WK211 left on 26 April, and the final jet, WK212, followed on 4 May. Pending its re-equipment with the Hawker Hunter, No. 56 Squadron, which never entirely completed conversion to the Swift, reverted to being wholly equipped with the Gloster Meteor F.Mk 8.

During its 14 months with the Swift, No. 56 Squadron had lost three F.Mk 1s in accidents, which had killed one pilot, and the F.Mk 1 and F.Mk 2 had been proven to be completely unfit for operational service. Some believed that with the

F.Mk 4, the aircraft was finally coming good, and Supermarine's Dave Morgan later insisted that, "The shortcomings of the aircraft were greatly exaggerated in order to support the cancellation of the Mk 4 fighter, which was a logical step in view of the reduction in world tension at the time." Certainly, with the end of the Korean war, the need for large numbers of fighters simply disappeared, while the money needed to develop the design and solve its problems could no longer be justified. It was much easier to concentrate funds on the Hunter and on development of the next generation of fighters.

Low level

Whether the condemnation of the Swift was fair or otherwise, the RAF had decided that the Swift was of no further use as a day fighter, and the fighter versions were abandoned. However, the aircraft had shown some excellent potential at low altitude, where it was fast and stable, and where the reheat system worked well. With its over-large fuselage, the Swift carried enough fuel to give the type a reasonable radius of action, and the decision was taken to press ahead with the planned tactical reconnaissance variant, the Swift FR.Mk 5.

The RAF's existing Meteor FR.Mk 9s were looking increasingly obsolete and structurally tired, while a reconnaissance version of the Hunter could not be produced for some years, without risking disruption and delay to the production of fighter variants. The Swift therefore seemed like an ideal interim tactical reconnaissance machine, at least in RAF Germany, though the Meteor was expected to soldier on in the Middle East and Far East.

RAF Germany then had two tactical reconnaissance squadrons, each equipped with the Meteor FR.Mk 9. No. II Squadron had moved from Bückeburg, to Gütersloh, then on to Wahn and finally, in 1955, to Geilenkirchen. No. 79 Squadron had received Meteors at Gütersloh in 1954, and had returned there after brief periods at Bückeburg, Laarbruch and Wunstorf.

The two squadrons followed slightly different concepts of operation, reflecting their different proximities to the inner German Border (with Communist East Germany). No. II Squadron tended to operate single aircraft, carrying underwing tanks, and often flying a hi-lo-hi sortie profile, while No. 79 flew shorter range missions with pairs of aircraft – often clean, and usually at low level from take-off to landing. With the Swift, both squadrons would switch to operating single aircraft at low level.

Reconnaissance prototype

One of the surviving F.Mk 1s (WK200) was converted to serve as the proof of concept aircraft for the FR.Mk 5, gaining an extended nose housing a fan of forward, left and right oblique F95 cameras. W. Vinten Ltd's F95 camera was the key to the Swift's success in the tactical reconnaissance role. Originally developed for the Meteor FR.Mk 9, and used by RAF Jaguars throughout the 1990s, the F95 framing camera gave hitherto unknown levels of sharpness and clarity even at very low level and high speed, and proved to be a cornerstone of the Swift's success in the reconnaissance role.

Though WK194 had undertaken trials with various air-to-ground weapons, including a battery of eight rockets under each wing, plans to arm the FR.Mk 5s with more than their built-in Aden cannon came to nothing, while the idea of a dedicated Swift fighter-bomber also faded.

While WK200 was essentially an F.Mk 1, the definitive FR.Mk 5 was a reconnaissance version of the F.Mk 4, with an afterburning Avon 114 engine, saw-tooth wing leading edges, a variable-incidence tailplane, and provision for a 220-Imp gal (1000-litre) ventral fuel tank. The aircraft also had an increased-height tail (compensating for the longer nose, but also seen on several of the F.Mk 4s) and a modern bubble canopy. Supermarine had unashamedly copied this latter feature from the Sabre. Unlike the F.Mk 4, the recce aircraft reverted to the twin-cannon arrangement of the F.Mk 1, though the wing – with its distinctive mid-span kink – was that of the F.Mk 2, albeit with an extended dogtooth leading edge and without fences.

Most of these features were tested on an F.Mk 4, WK272, which confirmed that the aircraft would comfortably exceed the 150-nm (278-km) radius of action demanded by the OR Branch. In fact, the Swift FR.Mk 5 was able to reach Berlin from Gütersloh, even after a reheat take-off and even with a five-minute dash at 545 kt (1009 km/h). This was modest for the Swift, which could attain 600 kt (1111 km/h) in reheat, or 565 kt (1046 km/h) in dry power, or 590 and 550 kt (1093 and 1018 km/h) with the ventral tank in place.

After the farcical simultaneous delivery of fighter Swifts to the AFDS and the front line, all before the A&AEE had finished testing, proper procedure was followed for the FR.Mk 5's intro-

Wearing No. II Squadron's triangle markings, a Swift FR.Mk 5 flies at low level along the Rhine. The Swift was admirably suited to the tac recce role, which did not expose its high-altitude shortcomings.

duction to service. An initial assessment was made by the A&AEE using WK272, after which 'A' Squadron evaluated XD903, the first true FR.Mk 5. The type was then exhaustively assessed by the AFDS at West Raynham. Sir Harry Broadhurst, then RAF Germany's AOC-in-C, then ordered his own evaluation, despatching one of his Meteor FR.Mk 9 squadron commanders to Boscombe Down to conduct his own assessment.

Mk 5 deliveries

The first four FR.Mk 5s went from Supermarine's South Marston factory to No. 23 MU on 21 December 1955, with Dave Morgan leading the formation. Supermarine's test pilots 'Chunky' Horne and 'Pee Wee' Judge were accompanied by Flight Lieutenant Norman Penney – a test pilot serving with No. 41 Group, which ran the RAF's Maintenance Units. The Swifts were delivered from Northern Ireland to RAF Germany by MU and squadron pilots, and by the RAF Benson-based Ferry Wing.

The Swift already had a relatively good range thanks to its internal volume, but this could be further increased by the carriage of a belly tank. Although it looked cumbersome, the belly tank only affected straight-line performance slightly.

While it was fast at low level the Swift FR.Mk 5 was not noted for its agility, although it could put on a good show in the right hands. The recce aircraft could shoot back thanks to its two 30-mm cannon under the intakes.

Supermarine Swift FR.Mk 5

Powerplant: Rolls-Royce Avon 114
Thrust: 7,175 lb (31.93 kN), 9,450 lb (42.05 kN) with reheat
Span: 32 ft 4 in (9.86 m), F.Mk 7 36 ft 1 in (11.00 m)
Length: 42 ft 3 in (12.88 m), F.Mk 7 44 ft 0½ in (13.42 m)
Height: 13 ft 2½ in (4.03 m)
Wing area: 347.9 sq ft (32.32 m²)
Basic weight: 13,735 lb (6230 kg)
Max landing weight: 17,000 lb (7711 kg)
Maximum take-off weight: 21,500 lb (9752 kg)
Fuel: 726 Imp gal (3300 litres), of which 506 Imp gal (2300 litres) carried internally
Maximum speed at low level: 608 kt (1126 km/h; 700 mph)
Maximum speed high level: Mach 0.92 in level flight
VNE: 600 kt (1111 km/h; 690 mph)
Limiting Mach no: 0.92 below 20,000 ft (6096 m), no limit above
Time-to-climb: 40,000 ft (12192 m) in less than 4.5 minutes with reheat
Range: 445 nm (824 km; 512 miles) at 'best range' speed (sea level, full fuel)
Ceiling: 45,800 ft (13960 m)
Load factor: +7.5 *g* (6.5 *g* with ventral tank)

No. II Squadron received its first Swift at Geilenkirchen on 23 February 1956, though the unit made a slow start, delayed by the failure of the Air Ministry to issue vital modification kits. No. 79 Squadron at Wunstorf received its first aircraft on 14 June 1956.

Both Swift squadrons provided a pilot for the 2 ATAF team for the May 1957 NATO Royal Flush II recce competition. The two officers came first and second in the individual placings, the Swifts won their class, and 2 ATAF seized the Gruenther Trophy. The Second Allied Tactical Air Force won Royal Flush III, in August 1958, but the Swift's performance was merely workmanlike, and the pilots (one from each unit) failed to top the scoreboard.

In March 1959, No.79 Squadron won the Sassoon Trophy, competed for by all RAF Germany reconnaissance squadrons, and this proved to be excellent preparation for Royal Flush IV in August 1959. The Swift team (with two pilots from No. 79 Squadron) won its class, and ensured another 2 ATAF victory. After another Sassoon competition, Royal Flush V in 1960 did not go so well for the Swift, and although one No. 79 Squadron pilot took the individual honours, 4 ATAF took back the Gruenther Trophy.

Away from the competitions, life on the Swift squadrons was action-packed and fun-filled, with plentiful flying and opportunities to sample the Hunter, as well as the Vampire and Meteor trainers maintained for instrument flying training and check flights. The combination of high spirits and very low level flying sometimes spilled over into 'licenced hooliganism', however. On one occasion, two Swift pilots narrowly escaped court martial after flying under a bridge on the Kiel canal. To their good luck, when the investigating officer visited the bridge he saw a formation of four Venoms do exactly the same thing – the problem was clearly too widespread to court martial all of the offenders!

Serviceability declined in 1958, as the Swift squadrons lost groundcrew to new Hunter units. By 1960, the situation was critical, and No. 79 Squadron was briefly declared non-operational.

Many pilots experienced both the Swift FR.Mk 5 and the Hunter and, surprisingly, many preferred the Supermarine aircraft as a pure flying machine, despite its marginal high-altitude performance and poor rate of turn. Many remarked on the excellent forward view, while others praised the less sensitive, very positive powered ailerons, the excellent (if heavy) manual controls, the stable ride at low level and the aircraft's confidence-inspiring ruggedness. "You felt as though you could knock down anything you had the misfortune to run into," one former Swift pilot observed.

As an operational aircraft, the Swift was faster at low level than its main competitors (the RF-84F and RF-101), and with two 30-mm cannon packed a heavier punch. It was also more agile, making it better suited to route searches, following a twisting road or river. Perhaps the best tribute to the Swift was paid by Fiat, which was clearly influenced by the Swift when it came to design the G91R, choosing the same twin Aden 30-mm cannon armament and the same camera fit of three F95s.

It has been claimed that the Swift was the best short-range, day-only armed tactical reconnaissance aircraft of its day, even including the Hunter FR.Mk 10 that succeeded it. It was certainly a very much more successful and useful aircraft than its fighter predecessor.

Vickers Supermarine Swift FR.Mk 5
No. IV Squadron
RAF Gütersloh, January 1961

This Swift is depicted as it appeared after No. 79 Squadron was renumbered as No. IV Squadron. No. 79's red arrow markings were retained, but No. IV's lightning flash markings were added to the nose. Within days of the renumbering the squadron received its first Hunter FR.Mk 10 to replace the Swift, a process completed by early February. The other RAFG recce squadron, No. II, completed its conversion in April

A close-up of the nose of an FR.Mk 5 shows the window for the side-facing F95 camera, and the undernose bulge mounted on the nosewheel bay door. The modern canopy gave the pilot a good view outside, especially to the sides.

Missileer

Though the proposed unarmed high-altitude PR.Mk 6 variant was abandoned at the same time as the F.Mks 1-4 because of the high-altitude reheat problems, the Type 550 did form the basis of the final production fighter version, the Swift F.Mk 7.

Originally, there had been plans for an operational force of Swift fighters armed with the Fairey Fireflash AAM, and Contract 9757 was issued for 75 of these Swift F.Mk 7s. These aircraft were to feature an extended nose, housing Ekco radar, and a longer span wing for better high-altitude performance, while the guns gave way to increased fuel tankage. These were all features of the PR.Mk 6, whose design therefore formed the basis of the new variant.

The 330-lb (150-kg) Fairey Fireflash (originally known as Blue Sky) was a beam-riding air-to-air missile intended for use against large, non-manoeuvring bomber targets. It was meant to be launched at the enemy aircraft's stern aspect, within a +/- 15° cone, from a range of about 6,000 ft/1830 m (with a maximum range of perhaps 10,000 ft/3050 m). The basic missile airframe (containing the warhead and fuse) was accelerated to about Mach 2.4 using two boosters (strapped above and below the missile) which were then explosively jettisoned about one and a half seconds later, passing above and below the launch aircraft, and leaving the missile to fly down the radar beam 'projected' by the carrier aircraft's radar. Extensive trials of the missile were undertaken, with firings being undertaken from Meteor NF.Mk 11s from 1951.

Though plans for an operational force of Swift F.Mk 7s were abandoned, the aircraft was still seen as an ideal trials vehicle for the new weapon, and the decision was taken to order two prototypes under Contract 9929, and to complete 12 of the aircraft originally ordered under Contract 9757.

A converted Swift F.Mk 4 acted as an aerodynamic prototype for the F.Mk 7, before the two prototypes (XF774 and XF778) began firing trials from RAF Valley in 1956. XF778 had cameras in the nose, and filmed XF774 which fired the missiles. Meanwhile, in November 1956, XF113 and XF114 were delivered to Boscombe Down to the Handling Squadron and the Aeroplane and Armament Experimental Establishment for clearance work and trials.

Trials at Valley

The remaining 12 F.Mk 7s were delivered to the Guided Weapons Development Squadron at RAF Valley from March 1957. The aircraft then needed to have specialised instrumentation installed – this requiring removal of the engine. Pilot conversion occupied the squadron until June, when the aircraft began flying 'tracking' sorties against manned Firefly targets operating from Llanbedr. Live firings against the Fireflies (now in unmanned configuration) began in October 1957, and the trials eventually encompassed three night

firings (against flares towed by Meteor drones) in March and April 1958. Live firings ceased in November 1958, and the aircraft were retired to No. 23 MU at Aldergrove in Northern Ireland.

The unit fired some 64 Fireflash missiles, achieving an 86 per cent success rate. The weapon would have been useful against bomber targets, but the rival Firestreak held greater promise, and the Fireflash and Swift F.Mk 7 were quietly retired. The first F.Mk 7 to be withdrawn went to No. 23 MU on 20 September 1958. Eight more followed on 21 November 1958, and the final aircraft joined its sisters on 20 December.

By the time the F.Mk 7s were retired, the RAF Germany Swifts were beginning to look tired. The

WK 293

~GRANT RACE~

No. II (Army Co-operation) Squadron has the distinction of operating the Swift for the longest period, flying the FR.Mk 5 from February 1956 to April 1961. This line-up at Geilenkirchen highlights the taller fin of the Mk 5, introduced to offset the effects of the longer nose.

integral wing tanks had always been very prone to leaks, and this had necessitated keeping the aircraft outside wherever possible, while National Service groundcrew had sometimes treated the aircraft carelessly or without finesse. The well-known Swift phenomenon of intake 'panting' (when the intake skin 'oil-canned' at between 6,500 and 7,000 rpm) had sometimes loosened rivets or even caused cracks that needed patching, while there were many instances of the engines ingesting foreign objects. The aircraft were also heavily used in the air, and by 1960 many aircraft required the replacement of their mainplane bolts and bushes.

There is little doubt that the RAF Germany Maintenance Unit (No. 420 R&SU) could have solved the problems and, indeed, there were many low-houred Swifts sitting at No. 23 MU (some of them with 'delivery mileage only') ready for issue, but there was little appetite to retain the Swifts and their associated support infrastructure in a shrinking Command within a dwindling RAF. Replacing the FR.Mk 5s with (slower) Hunter FR.Mk 10s offered compelling cost and logistics advantages, and (to the chagrin of some pilots) the decision was taken that the Swifts would be retired. There were also felt to be profound flight safety implications since, although only one front-line FR.Mk 5 pilot was killed (struck by his cockpit canopy when

it came off in flight), 10 FR.Mk 5s were written off in accidents, and many more made successful dead-stick landings. When the Directorate of Flight Safety investigated the Swift in 1958, it had concluded that the aircraft was unsatisfactory in safety engineering terms, with twice the technical major accident rate of the rival Hunter.

Mk 5 rundown

Rundown of the FR.Mk 5 force began in earnest on 20 October 1959, when 27 of the 32 aircraft in store at No. 23 MU were declared non-effective and put up for sale. XD912 had already been made surplus on 30 September 1958 and scrapped in January 1959, while WK274 was scrapped in May 1960.

The 10 former GWDS F.Mk 7s at Aldergrove were sold to W.H. Bushell for scrap on 2 February 1960, and the company bought single FR.Mk 5s on 27 February and 30 May, before buying 28 further surplus FR.Mk 5s on 27 June 1960. Four of these had never got further than No. 23 MU – never having been issued to the front-line. Bushell's yard at Aylesbury was briefly full of some 40 Swifts, but they were quickly broken down and processed.

This clearout left No. 23 MU with just two FR.Mk 5s, both of which had been delivered straight to storage, neither of which had ever flown with a service unit. They were sold to Lowton Metals of Warrington on 2 May 1961.

By then, the remaining Swifts in RAF Germany had either been scrapped on station, or flown back

to the UK for scrapping or for use on various stations' fire dumps. Twenty of the aircraft returned to No. 60 MU before being allocated to fire sections, where their lives were very short, while others flew directly to the stations where they would meet their fate. One Swift (WK295) saw brief service as Laarbruch's gate-guard, and two more ground instructional airframes escaped the attentions of the firemen and found their way into museums, WK277 ending up at Newark (having been purchased by Vickers Armstrong test pilot 'Dizzy' Addicott for conversion into a jet-powered car!), and WK281 passing to the RAF Museum's reserve collection at Colerne, later moving to Finningley, Swinderby and Hendon. The aircraft is now 'on loan' at Tangmere.

No Swift FR.Mk 5s escaped to fly on with test or trials units after the re-equipment of the RAF Germany squadrons, and only two Swifts remained airworthy by the summer of 1961. They were the two pre-production Swift F.Mk 7s, XF113 serving with the Empire Test Pilot's School at Farnborough, and XF114 flying wet runway braking trials for the Ministry of Supply. The ETPS aircraft was retired soon afterwards, but XF114 remained flying until 1967.

Fourteen F.Mk 7s were produced, dedicated to tests of the Fairey Fireflash beam-riding missile. The missile consisted of a central main body with warhead and fuze, and two booster rockets. The main campaign of tests was undertaken between October 1957 and November 1958, proving quite successful. However, the IR-guided Firestreak was adopted instead.

The last active Swift was F.Mk 7 XF114, seen here in 1968 just after its retirement from RAE service. The lack of mainwheel doors and the Donald Duck cartoon on the nose alluded to its last duties on wet runway tests.

This aircraft had been delivered to the A&AEE at Boscombe Down in March 1957, and was originally to have been used for familiarising Royal Navy pilots with the type. From September, the aircraft was used for various tests relating to the "coefficient of friction between aircraft tyres and wet runways" (aquaplaning). An initial series of trials was conducted at at Wisley, Pershore, Coltishall, West Raynham, Upper Heyford and Filton under contract 6/Acft/15373/CB9(c). These lasted until a second contract (KC/W/063/CB9)c) was awarded in February 1962, covering further trials at Wisley, Heathrow, Thurleigh, Weybridge and finally Cranfield. Painted black, with a caricature of Donald Duck skidding along on his bottom on the nose, the aircraft was stripped of its main undercarriage doors for the trials. The aircraft was finally struck off charge on 14 April 1967, and after a brief period in storage at Aston Down, the aircraft was sold to the Flint Technical College, (later Kelsterton College of Technology) as a ground instructional airframe.

New life for '114?

It briefly looked as though XF114 might be returned to airworthy condition, when Jet Heritage at Bournemouth swapped the aircraft for a Jet Provost. The Swift was moved to Hurn in January 1989, where Jet Heritage began a painstaking restoration of the aircraft. Unfortunately, Jet Heritage went into liquidation on 19 August 1999, by which time the Swift's fuselage was ready for electrics and engine to be installed, and the wings were being X-rayed by BAE Systems. The Swift moved to Scampton with Jonathan Whaley, Jet Heritage's former Managing Director, but little further progress was made before the aircraft was sold to Solent Sky, the former Southampton Hall of Aviation, in mid-2004. The aircraft remains in store, though there are plans to complete restoration to static display standards. Solent Sky also has a Swift nose section already on display.

There are a handful of other Swifts (or parts of Swifts) still extant. The nose of the former ETPS F.Mk 7 (XF113) is in the museum at Boscombe Down, and the rest of the hulk remains in a scrapyard near Frome. The fuselage of the historic, record-breaking WK198 was recovered from a scrapyard at Failsworth and has been restored by the North East Aircraft Museum, and a whole (but tatty) F.Mk 4, WK275 still sits outside an army surplus store in the village of Upper Hill near Leominster in Hereford & Worcester.

Jon Lake

Two Swift FR.Mk 5s are preserved for display, one kept inside at Tangmere in West Sussex, and this aircraft at the Newark Air Museum in Nottinghamshire.

Swift variants

Type 510

The Supermarine 510 was a research aircraft built to Specification E.38/46, intended to explore the flying characteristics of swept wings. Two were ordered, based on Attacker fuselages, and the first of these (VV106) made its maiden flight on 29 December 1948. It had 40° swept wings (with a span of 31 ft 8½ in/9.66 m) and tailplanes mated to an Attacker fuselage, and was powered by a 5,000-lb (22.25-kN) thrust Rolls-Royce Nene 2 engine. It was the first British aircraft with both swept wings and swept tailplanes.

The aircraft had leading-edge slats on the outer wings, but they were locked shut. Originally completed with a pointed nose and a blown canopy, the aircraft soon gained a more heavily framed canopy, and a blunter, standard Attacker nosecone.

The aircraft was briefly fitted with an A-type arrester hook, provision for up to four RATOG bottles and uprated (16-ft/4.9-m per second) main gear oleos for swept-wing carrier deck landing trials. After dummy deck trials at Farnborough, the Type 510 became the first swept-wing aircraft to make a carrier landing when Lt Jock Elliot landed aboard HMS *Illustrious* on 8 November 1950.

The planned second Type 510 (VV119) was modified during construction, emerging as the Type 528, which is described separately.

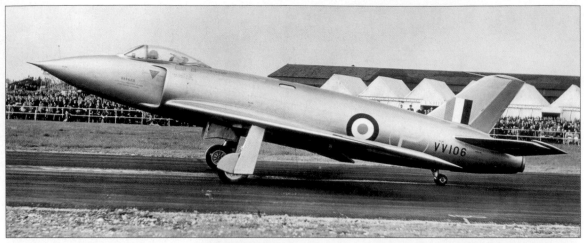

Above: The Type 510 is seen in its original configuration during the 1949 SBAC show at Farnborough. The nose and canopy were altered soon after.

Following a wheels-up landing on 14 December 1952, VV106 was rebuilt to a new standard, becoming the Type 517.

With the addition of a more rounded nosecone the lines of the Swift can be seen emerging from the original Type 510 design. Here the aircraft cruises over Poole Harbour.

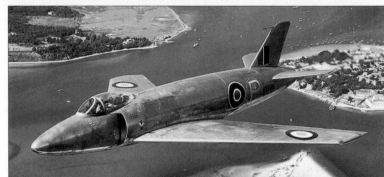

Type 510 (RAF) and Type 510 (Australian)

The Type 510 designation was briefly applied to proposed operational versions of the later Type 535, with four 20-mm cannon in the wings, leading-edge slats, and a tricycle undercarriage, and powered by a Nene or Tay engine. The type was examined by the RAF and the Royal Australian Air Force.

Type 517

In December 1948 Supermarine had proposed a number of improvements that it hoped would allow the Type 510 to reach Mach 1. In order to maintain control at transonic speeds, the aircraft would need a fully adjustable powered tailplane, with narrower chord elevators, while the company also sought to remove the elevator spring tab. In the event, the Type 510 was instead rebuilt with a new rear fuselage, all of which moved (with the tailplane) through a range of 8°. This had the same effect as

VV106 is seen on display after retirement, having been upgraded to Type 517 standard with an all-moving rear fuselage.

an all-moving tailplane, and proved effective and reliable, allowing the aircraft to operate at speeds of up to Mach 0.95.

After retirement on 14 January 1955 the aircraft became a ground instructional airframe at RAF Melksham, then at Halton, then became a gate-guard at Cardington, finally entering the RAF Museum at Colerne (and later moving to Cosford).

Type 520

The Type 520 designation was applied to a projected Avon-engined conversion of VV119, and also to a proposed operational version of the Type 510, with four 20-mm cannon in the wings and with outboard wing leading-edge slats.

Type 528

Originally laid down as the second Type 510, VV119 was completed with a revised rear fuselage, intended to allow installation of a 1500°K afterburner. Compared to the first 510, VV119 also had repositioned intakes (6 in/15 cm further forward), shallower but of 20 per cent greater area, and the 'operational wing' proposed for the Type 520, with provision for cannon and ammunition, and with reduced-span ailerons. Minor refinements to the undercarriage gave a reduced ground angle. The aircraft had the original pointed nose, but always had the more heavily framed canopy. As the Type 528, VV119 first flew on 27 March 1950, but went back into the experimental shop just over a month later, on 6 May, for conversion to Type 535 standards.

Type 531

The Type 531 designation was applied to a projected conversion of the Type 528 (VV119) with a Nene 3 engine, a variable-incidence tail, and nosewheel. Without the variable-incidence tail, the Type 535 designation was used.

Type 532

The Type 532 designation was applied to a projected conversion of the Type 510 (VV106) with a de Havilland Ghost engine.

Type 535

Ordered as a Type 510, then completed and flown as the Type 528, VV119 subsequently became the first member of the Swift family with a tricycle undercarriage, emerging after modifications as the Type 535. The aircraft's main undercarriage bays were moved aft of the mainspars, and a new nosewheel bay was added in the lengthened nose. The aircraft retained tailwheels, and could land on these to reduce the landing distance, pitching forward onto the nosewheel at 65 kt (120 km/h; 75 mph). The wing was of slightly increased area, with reduced sweep on the inboard trailing edge, and though armament was never fitted, the wing did briefly gain dummy cannon port fairings on the leading edge.

The aircraft made its first flight in its new guise on 23 August 1950, and made its first flight with the afterburner operational on 1 September, becoming the only Nene-engined aircraft to fly with an operational reheat system. The afterburner was later removed to save weight. The aircraft also briefly used perforated airbrakes on the upper surfaces of the wings, but they were removed, faired over, and replaced by use of a 35° flap setting (the landing flap setting was 60°).

The longer nose and tricycle undercarriage made the Type 535 a much better looking aircraft than its predecessors, and after the addition of a curving fillet between fin leading edge and fuselage it was photogenic enough for the film-makers to choose it to play the part of the fictional 'Prometheus' in the cinema film *Sound Barrier*.

The aircraft later gained a variable-

Three views show the Type 535 in various guises. Above is the original configuration (note deployed tailwheel) while below it is fitted with dummy cannon ports. At right it has been fitted with a curved fin fillet.

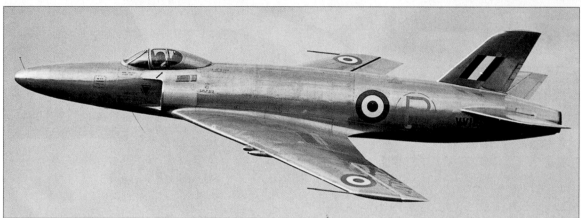

incidence tailplane (with a 13 per cent range of movement). When this had first been proposed, when the Type 528 was being converted, it was said that the designation would have changed to Type 531, but in the event, VV119 remained a Type 535 even with the new tailplane.

The Type 535 saw extensive service in the test and trials role, evaluating wing fences and anti-*g* clothing before flying with two and then four Fireflash AAMs from 28 May 1953. The aircraft later went to the Central Fighter Establishment for arrester barrier trials. The aircraft was grounded on 28 September 1955, and became a ground instructional airframe at Halton, after which it was unceremoniously scrapped.

Type 541 Swift

The Type 541 was the planned production model of the Type 535. The first two Type 541 aircraft acted as prototypes for the production Swift F.Mk 1, and were closely based on the Type 535 configuration.

In order to get the aircraft into service as quickly as possible, changes were kept to a minimum, though it was clear that the Rolls-Royce Nene engine of the Type 510, 517, 528 and 535 would be replaced by the more modern Rolls-Royce AJ.65 Avon, which offered about 50 per cent greater thrust, being rated at 7,500 lb (33.38 kN). Based on an axial compressor, rather than a centrifugal one, the Avon was much slimmer than the Nene, but it was considered that it was too late to redesign the fuselage so the Type 541 retained a much wider, fatter fuselage than was strictly necessary to house the new engine. The engine access doors above the fuselage were enlarged to improve access to the longer Avon engine.

The Type 541 looked sleek enough, but the fuselage could have been made thinner and still house the Avon engine.

WJ960, the prototype Swift, looked very similar to the Type 535 that preceded it, but switched to Avon power.

The armament was changed from four 20-mm Hispano cannon in the wings to two heavier 30-mm cannon in the forward fuselage, under the engine intakes. This allowed the wing to be redesigned, most notably with the ailerons being restored to their original span (the inboard edge had been moved outboard to make room for the guns on the Type 535). The wingtips were refined, leading to an 8-in (20-cm) increase in span to 32 ft 4 in (9.86 m). The flaps were simplified, though their appearance was unchanged.

A pitot head was added to the top of the tailfin, and the design of the canopy framing was refined. Tailwheels were initially retained, and doors (omitted from the Type 535) were added to the tailwheel bay. Later, the tailwheel assembly was removed altogether.

Two Type 541 prototypes were built, and these differed in detail. WJ960, which first flew on 1 August 1951, retained spring tabs on the ailerons, and therefore suffered from aileron flutter that limited it to subsonic speeds. It was so similar to the Type 535 that

Supermarine Test Pilot Dave Morgan frequently referred to the aircraft as the 'second Type 535' or as the 'Avon 535'.

The second Type 541 prototype (WJ965) first flew on 25 August 1952 and was more deserving of a new Type number. The aircraft differed from the first Type 541 prototype in having a revised fuselage, with a rather shorter tailcone, a modified nose, a production-representative tailfin and cockpit canopy, and the wing was repositioned, so that the aircraft more closely resembled the production Swift F.Mk 1s that followed.

WJ965 soon gained geared aileron tabs, and this allowed the aircraft to 'break the sound barrier', in a dive, doing so for the first time on 26 February 1953, over Chilbolton. WJ960 later gained a variable incidence tailplane and geared aileron tabs, and flew on until finally grounded on 15 September 1959.

With its shorter nose and repositioned wing the second Type 541 was a more representative prototype for the production aircraft that followed than the first machine.

Type 541 Swift F.Mk 1

The first Swift production variant was the F.Mk 1 day fighter, some 20 of which were built under Contract 5969 (WK194-213). The first two were built at Hursley Park, but the remainder were manufactured at South Marston. The first F.Mk 1 (WK194) made its maiden flight on 25 August 1952.

Nine F.Mk 1s entered front-line service from February 1954, becoming the RAF's first swept-wing fighter aircraft in the process. Two were lost in service. Others tested features of other, later marks or served as prototypes for later versions, WK198 serving as the F.Mk 4 prototype, WK200 featuring the

camera nose of the FR.Mk 5, and WK201 testing the four-cannon installation and ventral fuel tank developed for the F.Mk 2.

The F.Mk 1 was fitted with two 30-mm Aden cannon, had a conventional fixed tailplane and was

powered by a 7,500-lb (33.38-kN) thrust Rolls-Royce Avon 108 engine without reheat. After the type's withdrawal from No. 56 Squadron the survivors were allocated to the Schools of Technical Training at Halton (No. 1 SoTT), St Athan (No. 4 SoTT), and Weeton (No. 8 SoTT).

Two views show WK194, the first Swift F.Mk 1. Although it was notionally the first 'production' aircraft, it and the second aircraft were not built at the main South Marston works (today home to a Honda automobile factory).

Type 541 Swift F.Mk 2 and F.Mk 3

The second Swift variant was the F.Mk 2, which was essentially an F.Mk 1 with two extra Aden 30-mm cannon, bringing the total to four. The wingroots were redesigned and the inboard leading edges were extended to give extra room for the increased ammunition load. The extended inboard leading edge resulted in a leading-edge 'kink' at about mid-span, and this caused pitch-up problems under g at speeds above Mach 0.85, when the aircraft could depart quite violently.

The first F.Mk 2 (WK214) first flew in December 1952, though F.Mk 1s WK199 and WK201 served as prototypes. Apart from the kinked wing leading edge, the F.Mk 2s delivered to the RAF differed from the F.Mk 1s in having their topsides painted in Fighter Command's new two-tone disruptive grey and green camouflage. The effectiveness of this was marred, somewhat, by the large red, white and blue roundels and fin flashes then in vogue, and many aircraft gained No. 56 Squadron's famous red and white

The Mks 2 and 3 could be differentiated from the Mk 1 by having a slightly kinked wing leading edge. The Mk 3, as here, had wing fences fitted to improve the Swift's unpredictable handling at high speed.

checkerboards flanking the roundel on the rear fuselage.

Sixteen Swift F.Mk 2s were built to Contract 5969, and six of these were delivered to No. 56 Squadron from 30 August 1954. The pitch-up problem was ameliorated by the addition of ballast in the nose, though this had a detrimental effect on performance.

After the re-equipment of No. 56 Squadron, the surviving Swift F.Mk 2s were mostly allocated to the Schools of Technical Training at Halton (No. 1 SoTT) and St Athan (No. 4 SoTT), and to the RAF Technical College at Henlow.

The third Swift variant, the F.Mk 3 was similar to the F.Mk 2 but with the reheated Rolls-Royce Avon 114 engine originally tested on WK195. The F.Mk 3s

also gained overwing fences slightly further outboard than the leading-edge 'kink', and subsequently gained vortex generators above and below the tailplanes, helping to counter the pitch-up phenomenon. The 25 aircraft built to Contract 5969, were intended to have formed a second Swift squadron but, in the event, only three of the aircraft were used by Controller (Aircraft) and

the A&AEE, and the remainder were delivered (by road) straight into storage with No. 15 MU at Wroughton. None was issued to operational units, most becoming ground instructional airframes with the Schools of Technical Training at Halton (No. 1 SoTT), St Athan (No. 4 SoTT), Weeton (No. 8 SoTT), Kirkham (No. 10 SoTT), and Melksham (No. 12 SoTT).

Type 545

The Type 545 was designed as a private-venture supersonic version of the Swift, and was launched following the early success of the Type 510, before the definitive Swift's problems had surfaced. Two designs were drawn up, one with a 50° swept wing, the other with a new high-aspect ratio cranked wing. Both designs were drawn up around a RA.12R engine.

Supermarine's second Type 545 (known as the Type 545 Stage 2) proposal promised greater performance advantages over the standard Swift, and was accepted by the Ministry of Supply in March 1951. A formal order for two prototypes was placed in February 1952, under Specification F.105D2, officially issued on 14 October. Service entry was set for 1957. The rival Hawker P.1083 was cancelled in July 1953.

The aircraft was an all-new design, albeit with clear 'Swift' ancestry, and combined an area-ruled fuselage with a compound-sweep 'crescent wing', and other aerodynamic and structural refinements. The aircraft had a

bifurcated nose intake with a radar in the centrebody.

The first prototype (XA181) had a Rolls-Royce Avon RA.14R, rated at 9,500 lb (42.28 kN) thrust or 14,500 lb (64.53 kN) with afterburning. It was expected to attain Mach 1.3 in level flight. It was to have flown during the spring of 1953, but this was delayed after all-weather capability with AI radar was added to the specification, and after it became clear that the aircraft would only be able to attain its design speed in a dive, when the project was placed under review.

It soon became obvious that any production Type 545 would need a larger fuselage and a major redesign, and that the prototypes already under construction would be useful for little more than aerodynamic trials.

It was briefly hoped that a development batch of operational aircraft, powered by RA.23R engines, would enter service in 1958, and that these would have a Mach 1.3 capability and a ceiling of 51,000 ft (15545 m), before the definitive RB.106-powered aircraft arrived two years later, with a 60,000-ft (18288-m) ceiling and Mach

1.7 or Mach 2.0 capability.

The scheme soon fell into difficulties – the RAF's OR.329/F.155T requirement called for a two-seat interceptor, while the Gloster GA.6 Thin Wing Javelin and English Electric P.1 both offered similar or better performance with lower technical risk. As wind tunnel testing raised further doubts about the Type 545, enthusiasm for the programme waned. Work on the first aircraft was therefore halted, while the second was scrapped.

A new specification, E.7/54, was issued to allow XA181 to be completed for trials use by the RAE, with a revised first flight date in July 1955, and with a new RA.28R engine, rated at 10,150 lb (45.17 kN) thrust or 14,000-lb (62.3 kN) with afterburner. Work on the prototype ceased on 23 August 1955, about three months before flight trials were to begin, and the whole project was cancelled on 22 December. The airframe (virtually complete) was transferred to the Cranfield Institute of Technology, where it served as a ground instructional airframe until scrapped during the mid-1960s.

The second prototype (XA186) was

to have been finished to a revised design, more representative of the planned operational configuration, with a longer nose and a wider centrebody, and with the mainwheels moved back by one fuselage frame. The aircraft was to have been powered by an RA.24R Avon, rated at 11,250 lb (50.06 kN) thrust or 14,350 lb (63.86 kN) with afterburner. The outboard underwing pylons were to have been 'plumbed' for the carriage of underwing fuel tanks.

The operational aircraft would have had a shorter nose than either prototype, with an 18-in (46-cm) conical radome above a Crusader-type chin intake (resembling the nose of the F-86D). The aircraft was to have had a rearward-retracting nosewheel, and a rather shallower canopy and windscreen, while the rear fuselage was rather fatter to accommodate the larger diameter, fully-variable convergent/ divergent afterburner. The aircraft was to have been powered by an RA.35R (rated at 11,000 lb/48.95 kN or 16,300 lb/72.54 kN with afterburning) or an RB.106 engine (15,000-lb/66.75 kN dry). The latter engine was expected to give Mach 1.68 performance.

Type 546 Swift F.Mk 4

The next production Swift variant was the F.Mk 4, which added a variable-incidence tailplane to try to solve the pitch-up problems of the F.Mk 2 and F.Mk 3. The new tail did fix the problem to an extent, but the fact that reheat would not light at high altitude fatally damaged the new variant's operational effectiveness, and production was cut back.

Some 39 Type 546 Swift F.Mk 4s were to have comprised the balance of the initial 100-aircraft order, augmenting 20 F.Mk 1s, 16 F.Mk 2s and 25 F.Mk 3s. Of an initial contracted batch of 39, only eight were completed and flown as F.Mk 4s (WK272-279), the balance being converted to FR.Mk 5 standards before delivery. No F.Mk 4s were delivered to the RAF, the aircraft remaining in the hands of Controller (Aircraft) for development and trials use until retired or converted. Four of the aircraft flown as F.Mk 4s (WK274, 276, 277 and 278) were soon converted to FR.Mk 5 standards.

Because they were used for test and trials duties, several of the F.Mk 4s incorporated features of later marks. WK275, for example, had the taller tailfin, dogtooth leading edge and blown canopy of the FR.Mk 5, as well as a unique single-piece slab tail.

The eight production F.Mk 4s were augmented by the fifth F.Mk 1, WK198, which served as the F.Mk 4 prototype, flying for the first time on 27 May 1953. This aircraft was subsequently used for a number of record-breaking flights, smashing the London-Paris record in May 1953, before the same pilot, Mike Lithgow, used the same aircraft to break the world air speed record on 26 September 1953, clocking an impressive 735.4 mph (1183.5 km/h).

The best-known of the Swifts, the F.Mk 4 prototype achieved fame as the world speed record-holder, although its reign as the world's fastest aircraft lasted for just one day.

Right: The F.Mk 4 prototype began life as the fifth production F.Mk 1, but was retained for trials of the variable-incidence tailplane that distinguished the Mk 4. It inherited the kinked leading edge and overwing fences from the Mk 3. It initially flew with a nose probe, as seen here, but was later fitted with a standard nosecone.

Type 547a Swift T.Mk 8

The Type 547 designation was applied to studies for a two-seat version of the Swift with a side-by-side cockpit layout, like that of the rival Hunter trainer. The Type 547a operational conversion trainer would have used the designation T.Mk 8.

Type 547b Swift night-fighter

The two-seat design studies included the Type 547b, which was an operational all-weather fighter, based on the trainer, but with an extended radar nose and with provision for two or four Aden cannon.

Type 548 Naval Swift F.Mk 4

Though Fleet Air Arm interest was primarily focused on the twin-engined Scimitar, there were a number of studies for navalised Swift variants. Of these only the Type 548 (a navalised F.Mk 4) received a Supermarine type number. The aircraft would have been equipped with an arrester hook, and 20 were ordered in March 1952 to meet specifications N.105D and NR/A.34, but this order was cancelled later that year.

Type 549 Swift FR.Mk 5

With the failure of the fighter Swift, the decision was taken to adapt the type for use in the low-level tactical reconnaissance role, to replace the Meteor FR.Mk 9. The resulting Type 549 Swift FR.Mk 5 was based on the F.Mk 4, with the same afterburning engine. Since the type was intended to operate at low level, the reheat problems experienced at high altitude were irrelevant.

Fifty-eight FR.Mk 5s (XD903-930, 948-977) were ordered under contract 9463 (the last was never completed). They were followed by 39 aircraft

The FR.Mk 5 could be easily identified thanks to its lengthened nose and outer panel leading-edge extensions. Even with the improvements the Swift was never in the class of the Hunter, although it was a useful tactical recce platform.

(WK280-WK315) originally ordered as F.Mk 4s under Contract 5969, a single new-build aircraft (WN124, also part of Contract 5969) and four conversions from F.Mk 4s (WK274, 276, 277 and 278). This gave a total of 102 FR.Mk 5s. Thirty-one further new-build FR.Mk 5s were cancelled.

The FR.Mk 5 was designed with a longer nose with a fan of three new

F.95 cameras and reverted to the F.Mk 1's armament scheme of two 30-mm Aden cannon. The wing was redesigned, incorporating a saw-tooth leading edge in place of the fences

used on previous variants. Provision was made for the 220-Imp gal (1000-litre) ventral fuel tank tested on a number of F.Mk 1s. Finally, the FR.Mk 5 replaced the fighter Swift's heavily-

framed canopy with a more modern 'bubble' canopy, reminiscent of that fitted to the Attacker and the original Type 510.

An F.Mk 1 (WK200) was fitted with

the extended camera nose planned for the FR.Mk 5 (and with the taller tailfin used by some F.Mk 4s), and acted as a prototype for the production reconnaissance variant. This aircraft lacked the definitive sawtooth leading edge and clear view canopy, however, and retained Fighter Command's original 'high-speed silver' finish. The first true FR.Mk 5, XD903, made its maiden flight on 25 May 1955.

The FR.Mk 5 entered service with the RAF in February 1956, equipping Nos II and 79 Squadrons in RAF Germany. The type served until replaced by the Hawker Hunter FR.Mk 10 from 1960.

Though the Hawker Hunter was considered to be one of the best fighter reconnaissance aircraft of its time, many pilots missed the Swift's afterburner-aided straight-line performance, as well as the aircraft's reassuringly strong and rugged airframe.

Type 550 Swift PR.Mk 6

The Type 550 Swift PR.Mk 6 was intended as an unarmed strategic reconnaisance aircraft to replace the Meteor PR.Mk 10 in the high-altitude reconnaissance role. The aircraft was based on the FR.Mk 5, but had a

modest increase in wingspan and a lengthened nose, with provision for extra vertical survey and mapping cameras. Problems with the reheat on the Avon RA.7 at altitude led to the project being abandoned. The PR.Mk 6

prototype, XD943, was almost complete when the programme was cancelled in April 1954. The airframe was complete enough to be useful as a ground instructional airframe, and the aircraft was sent to RAF Halton.

Type 551

The Type 551 designation was applied to WK199, an F.Mk 2. The aircraft was supposed to have been re-engined with an Avon 105, though it is unclear whether this happened.

Type 552 Swift F.Mk 7

The last production variant was the Type 552 F.Mk 7. This was designed as the first Swift variant to be armed with guided air-to-air missiles, and was originally envisaged as an operational interceptor, powered by a more powerful Rolls-Royce Avon 116 engine, rated at 7,350 lb (32.71 kN) thrust or 9,950 lb (44.28 kN) with afterburning. The operational aircraft was to have provision for four beam-riding Blue Sky

AAMs underwing and four Aden cannon in the fuselage, and with Ekco radar guidance equipment in the extended nose originally designed for the PR.Mk 6. The wingspan was extended by some three feet.

Contract 9757 originally covered 100 aircraft, but in the event, the 14 F.Mk 7s were actually built as dedicated guided-missile trials aircraft, for firing trials of the Fairey Fireflash air-to-air missile (as

Blue Sky had become). Cannon armament was deleted from these F.Mk 7s, making way for increased internal fuel tankage (bringing the total to 548 Imp gal/2491 litres from the FR.Mk 5's 506 Imp gal/2300 litres), while the aircraft was designed to use AVTUR in place of AVTAG.

Some 11 of the F.Mk 7s (XF113-XF123) were built under Contract 9757, two prototypes (XF774 and XF778) under Contract 9929 and the other (XF124) under Contract 8812.

The new F.Mk 7s were preceded by a partially converted F.Mk 4, WK279, which carried Blue Sky missiles and was aerodynamically representative of the F.Mk 7, before the true prototypes (XF774 and 778) began firing trials in 1956. For these early trials, XF778 had cameras in the nose, and filmed XF774 that fired and guided the missiles.

Type 554

The Type 554 designation was applied to a stillborn trainer version of the Type

545 designed to meet OR318, with side-by-side seats under a clamshell canopy. The trainer, which was to have been powered by an RA.19R Avon

engine with a fully-variable afterburner, retained an 18-in (46-cm) radar scanner in the upper nose, and had provision for underwing missiles, but the internal

armament was reduced to a single Aden cannon on the port side, with 200 rounds of ammunition.

Swift operators

Before the failure of the Swift fighter, and subsequent cancellation of many contracts, orders for the Swift reached 492. One hundred had been ordered in November 1950, 50 more in January 1951, 140 in September 1952, 110 in June 1953 and 75 in August 1953. Seventeen further aircraft were ordered for the Ministry of Supply.

In the event, Supermarine built 184 Swifts, comprising two Type 541 prototypes, 20 F.Mk 1s, 16 F.Mk 2s, 25 F.Mk 3s, eight F.Mk 4s, 98 FR.Mk 5s, one PR.Mk 6 and 14 F.Mk 7s. Of these, just nine F.Mk 1s, six F.Mk 2s and 70 FR.Mk 5s entered service with front-line units, 10 F.Mk 7s served with the GWDS, while a few more served with the A&AEE and second-line units like the Central Fighter Establishment. When the Comptroller and Auditor General, Sir Frank Tribe, reported on the Swift programme in 1956, his assessment that 129 flyable Swifts (28 F.Mk 1s and 2s, 101 FR.Mk 5s and F.Mk 7s) had been delivered to the RAF (at a then-staggering cost of £40 million) was over-generous.

His assessment of the number of aircraft actually built was more accurate – with a total of 170 aircraft broken down into 152 for the RAF (including 23 F.Mk 3s delivered straight into storage) and 18 for the Ministry of Supply.

These aircraft equipped a single Fighter Command squadron and two reconnaissance squadrons in RAF Germany (though these reconnaissance units used three different numberplates).

The RAF's Maintenance Units and Schools of Technical Training played a disproportionately important part in the Swift story, since many aircraft never reached operational units, and were either stored or used as ground instructional airframes from whom they were accepted until they were scrapped.

Technical Training establishments using Swifts included No. 1 SoTT Halton, No. 4 SoTT St Athan, No. 8 SoTT Weeton, No. 10 SoTT Kirkham, No. 12 SoTT Melksham and the RAF Technical College at Henlow. Maintenance Units storing, servicing or dealing with Swifts included Nos. 15 MU at Wroughton, No. 19 MU at St Athan, No. 23 MU at Aldergrove, No. 33 MU at Lyneham, No. 38 MU at Llandow, No. 49 MU at Gatwick, No. 58 MU at Sutton Bridge, No. 60 MU at Church Fenton and No. 71 MU at Bicester, as well as No. 420 AS&RU in RAF Germany.

No. II (AC) Squadron
No. II (Army Co-operation) Squadron had a long and proud history in the reconnaissance role, settling in RAF Germany and re-equipping with the Meteor FR.Mk 9 in 1950 at Bückeburg. The squadron moved to Gütersloh in 1952, then on to Wahn in 1953 before relocating to Geilenkirchen in 1955. No. II Squadron received its first Swift at Geilenkirchen on 23 February 1956, though the unit made a slow start, delayed by a failure to issue vital modification kits. No. II Squadron moved to Jever in November 1957, joining the Hunter-equipped Nos 4 and 93 Squadrons, who operated in the day fighter role. No. II Squadron received its first Hunters in January 1961, and operated both types until 13 April 1961, when the last Swift was flown out to RAF Manston.

No. II Squadron's Swifts wore black bars, each with a white triangle superimposed, on each side of the fuselage roundel, with individual 'swept back' code letters in white or yellow high on the tailfin.

No. IV Squadron
Formed from the disbanding No. 79 Squadron on 1 January 1961, No. IV Squadron augmented its Swifts with a growing number of Hunter FR.Mk 10s, the first of these arriving on 5 January. The last Swift was flown out to RAF Church Fenton on 7 February.

No. IV Squadron's Swifts retained No. 79 Squadron's 'arrowhead' markings flanking the roundel, but several (WK293) had No. 4 Squadron's black and red 'sun' insignia added to the nose, flanked by black and red bars. Yellow lightning flashes divided the colours on the 'sun' and bars, which were thinly outlined in yellow.

No. 56 Squadron
One of the RAF's most senior fighter squadrons, No. 56 received its first Swift at Waterbeach on 20 February 1954. The squadron's 'B' Flight remained equipped with the Meteor F.Mk 8 throughout the Swift era, and 'A' Flight failed to complete conversion to the new aircraft. Operating under an extremely restrictive release, the Swifts never became operational, and a cease flying order was received on 15 March 1955.

Many of No. 56 Squadron's aircraft received the unit's distinctive red and white checkerboard markings (outlined thinly in blue) on each side of the fuselage roundel. Some aircraft carried an individual letter code on the fin, often 'swept', usually in red, and sometimes outlined in white.

No. 79 Squadron
No.79 had received Meteors at Gütersloh in 1951, and had returned there after brief periods at Bückeburg, Laarbruch and Wunstorf. No. 79 Squadron received its first aircraft at Wunstorf on 14 June 1956. The squadron moved to Gütersloh in August 1956, where it disbanded on 30 December 1960, though its personnel and aircraft then took over No. IV Squadron's numberplate.

No. 79 Squadron's aircraft wore white bars flanking the roundel, each with a forward-pointing red arrow superimposed. Code letters were unswept, and applied in yellow immediately above the fin flash. Some aircraft wore the squadron's Salamander badge on their engine intakes.

Ferry Wing
The Ferry Wing formed at RAF Benson in 1953. The Wing consisted of Nos 147 and 167 Squadrons, and was responsible for ferrying aircraft to RAF units worldwide, including RAF Germany, and the Near, Middle and Far East Air Forces.

No. 147 Squadron, which reformed on 1 February 1953 by redesignating No. 1 Long Range Ferry Unit, was mainly responsible for ferrying jet fighters and, as such, ferried most of the Swift FR.Mk 5s from South Marston to Aldergrove, and from Aldergrove to Germany. It was amalgamated with No. 167 Squadron on 15 September 1958 into a single Ferry Squadron. This disbanded in 1960 when front-line squadrons gained the responsibility for collecting and ferrying their own aircraft.

Aeroplane & Armament Experimental Establishment
Formed at Martlesham Heath on 24 March 1924 by renaming the former Aeroplane Experimental Establishment, the A&AEE's primary role was testing aircraft and weapons to ensure that they met the RAF's requirement, and were safe and fit for operational use. The A&AEE was responsible for issuing a release to service, and for setting any limitations. Aircraft were supposed to be evaluated by the A&AEE once manufacturer's testing was complete, and before they were delivered to service units, though this did not happen in the case of the early Swift fighter variants.

The A&AEE moved to Boscombe Down in 1940. By the 1950s, fixed-wing fighter aircraft were evaluated by 'A' Squadron, which briefly operated a succession of Swift variants. None of these wore unit markings.

Central Fighter Establishment
The Central Fighter Establishment (CFE) was the forerunner of modern operational evaluation units and was responsible for developing tactics and doctrine for new aircraft types, and to test them (as required) to ensure their operational suitability after they had been cleared for service by the A&AEE. The Central Fighter Establishment parented a number of specialised training, trials and evaluation units. CFE units operating the Swift included the Air Fighting Development Squadron and the Guided Weapons Development Squadron.

Air Fighting Development Squadron
The Air Fighting Development Squadron (AFDS) was tasked with the development of fighter tactics, and with evaluating new types, and the unit briefly tested and reported on the Swift F.Mk 1 and FR.Mk 5. AFDS pilots were often invited to fly new aircraft by the manufacturers (the CO, Wing Commander Bird Wilson flew a production Swift from Chilbolton on 1 October 1953, before the squadron took delivery of three F.Mk 1s of its own (WK201, WK202 and WK205) in February 1954. The squadron evaluated the FR.Mk 5 from June 1955.

Guided Weapons Development Squadron
The Guided Weapons Development Squadron stood up at Valley in March 1957, under the command of Wing Commander J.O. Dalley. Though it reported to the Central Fighter Establishment at West Raynham, the GWDS operated autonomously. Tasked with testing and evaluating the Fairey Fireflash missile, the unit took delivery of 10 Swift F.Mk 7s, which remained active until November 1958. The last Swift was retired on 20 December 1958 and, re-named as No. 1 Guided Weapons Training Squadron, the unit re-equipped with Gloster Javelins for trials with the IR-homing Firestreak missile. No unit markings or codes were applied to the unit's Swifts.

Empire Test Pilots' School
The Empire Test Pilots' School was one of a series of Empire flying schools founded during the war and staffed by personnel from all over the Empire. The Empire Test Pilots' School (ETPS) was founded in 1943 at Boscombe Down, moving to Cranfield in 1945, before settling at Farnborough in 1947. Tasked with training test pilots from all over the world to test all types of aircraft, the ETPS used a wide variety of aircraft, aiming to give students experience of types that would be unfamiliar. Aircraft which might be judged unsuitable for a particular role were always especially prized. The ETPS used one Swift F.Mk 7 from 1958-1962, the aircraft providing students with a reheat-equipped platform when that was still quite unusual.

The ETPS Swift F.Mk 7 wore standard camouflage, but used the individual code '19' on the rear fuselage.

Royal Australian Air Force
At least two Swifts (F.Mk 1 WK199 and F.Mk 2 WK215) were transferred from the RAF to the RAAF. It is possible that the Swifts were originally intended for evaluation purposes, though this seems unlikely. Both were used at Woomera as part of Operation Buffalo, providing targets to be analysed following exposure to an atomic blast. The aircraft were augmented by a number of retired RAAF Mustang airframes, and were probably buried after the tests were complete. Four further F.Mk 2s (WK216, WK217, WK221 and WK239) were allocated for Operation Buffalo in January 1956. Some reports suggest that these aircraft were not sent to Australia and were instead scrapped in June 1957.

No. 79 Squadron flew the Swift FR.Mk 5 from 1956 to 1960, when it renumbered as No. 1V Squadron. RAF Germany Swifts were all painted in standard RAF camouflage, with high-visibility roundels and squadron markings.

Junkers Ju 88

The Luftwaffe's versatile bomber

The Junkers Ju 88 – in its various guises – was the backbone of Luftwaffe striking power for most of World War II. More than 14,000 Ju 88s were built, and the type remained in large-scale production from the beginning of World War II almost until its end. The Ju 88 served with distinction as a bomber, as a long-range reconnaissance aircraft, as a long-range fighter and as a night fighter.

The aircraft that would become the Junkers Ju 88 began life in 1934, when the Luftwaffe issued a requirement for a high-speed, general-purpose combat aircraft to serve in the roles of bomber, long-range fighter and reconnaissance platform. In the months to follow the Luftwaffe specification shifted towards its requirement for a high-speed bomber. This aircraft was to carry a maximum bomb load of 800 kg (1,764 lb), attain a maximum speed of 500 km/h (311 mph), cruise at 450 km/h (280 mph) and attain an altitude

Wearing the yellow fuselage band that signified assignment to the Russian front, this Ju 88A served with 9. Staffel/Kampfgeschwader 51. The Ju 88 saw widespread use in Russia, including low-level attacks.

of 7000 m (22,965 ft) carrying its normal bomb load.

Henschel and Junkers produced competing paper designs to meet the requirement, and the Luftwaffe ordered the construction of prototypes of both types. The Junkers submission was designated the Junkers Ju 88, its competitor was the Henschel Hs 127.

Construction of the Ju 88 prototype began at the Junkers plant at Bernburg in May 1936, and it was completed in the following December. On 21 December 1936 the aircraft made its maiden flight with the company's chief test pilot, Flugkapitän Kindermann, at the controls. The aircraft incorporated several features then considered modern, including all-metal stressed-skin construction, flush riveting, electrically operated variable-pitch airscrews and undercarriage, and low-drag annular radiators for the 1,000-hp (746-kW) Daimler Benz DB 600Aa liquid-cooled engines.

Soon after it began testing the Ju 88 V1 prototype was lost in an accident, but in April 1937 flight-testing resumed with the V2. The next prototype, the V3, appeared in the following September powered by Jumo 211 engines. The first Ju 88 to carry defensive armament, it had the rear of the cockpit canopy raised to provide the single dorsal MG 15 7.9-mm machine-gun with a better field of fire.

Plagued by delays, the competing Hs 127 design made its maiden flight only late in 1937. That was too late and it left the field to the Ju 88, which was chosen as the winner of the fast bomber programme.

Following the disappointing results from Luftwaffe horizontal bombing attacks during the Spanish Civil War, that service revised the specification for its new bomber type. The aircraft had now to be strengthened and equipped to carry out dive-bombing attacks. The German predilection for this type

Above: This view highlights the position of the twin MG 81 guns installed in the rear of the canopy.

Opposite page: A Ju 88A runs up for a mission somewhere in the Mediterranean theatre. Note the bombs carried beneath the inner wings.

Below: In the later part of the war the Ju 88's most important role was night fighting. This is a Ju 88G-6 with an SN-2 radar array.

Junkers Ju 88 prototypes

The Ju 88 V1 climbs out during a very early test flight (above). It subsequently acquired the civil registration D-AQEN (left) before being lost during high-speed trials. The fairing above the canopy housed an aft-facing camera.

Below: The V4 was the first with the distinctive 'beetle's eye' glazing. It had a four-man crew and was stressed for dive-bombing. It first flew on 2 February 1938.

Above: The V3 was the first Ju 88 with Jumo 211 power (albeit early 211As), and the first with raised canopy glazing, although it retained the original nose shape. The blister under the nose housed a sight for bomb-aiming. It first flew on 13 September 1937.

Left: The V5 was a one-off 'special' for record-breaking purposes. It had a streamlined nosecone, lowered canopy and no ventral gondola. Powered by Jumo 211B-1s, it first flew on 13 April 1938.

Above: The V6 was the production prototype for the Ju 88A, first flying on 18 June 1938. It was followed by four further A-series prototypes that tested elements of the bomber's design, such as divebrakes and underwing pylons.

Below: This is the eighth of 10 Ju 88A-0 pre-production aircraft, distinguished from later machines by their four-bladed propellers. They served with Erprobungsstelle 88 at Rechlin on service trials.

of attack has often been criticised, but at the time dive-bombing was by far the most accurate means of delivering bombs against a defended target. It was, moreover, the only method to promise reasonable success against warships manoeuvring at speed in open water. As was said at the time, what was the point of carrying bombs to a target if the inaccurate mode of attack meant that few of them would hit it?

The V4 was the first Ju 88 stressed for the dive-bombing manoeuvre. Other new features included the 'beetle's eye' nose made up of optically flat glass panels, and a ventral cupola with a second rifle-calibre machine-gun firing rearwards.

During the 1930s the German National Socialist government devoted considerable effort to enhancing national prestige at every opportunity. One area it amply supported was that of building and modifying aircraft for record-breaking flights. To that end the Ju 88 V5 was fitted with two Jumo 211B-1 engines each rated at 895 kW (1,200 hp). The airframe was cleaned up, the ventral cupola removed, the rear canopy lowered and a more streamlined and pointed nose section fitted. Thus modified, in March 1939 the Ju 88 V5 broke the world record for the 1000-km (621-mile) closed circuit with an average speed of 516.6 km/h (321 mph) carrying a 2000-kg (4,409-lb) payload. Four months later the V5 broke another speed, distance and load record when it covered 2000 km (1,243 miles) at an average speed of 500 km/h (311 mph), with the same payload.

During the autumn of 1937 the Ju 88 was ordered into large-scale production. The V6, which first flew in June 1938, was the production prototype for the A series aircraft. Like the V5, it was powered by Jumo 211B-1 engines. Defensive arma-

Junkers Ju 88 production

Initially Ju 88 production was slow in getting into its stride, and only about a hundred of these aircraft were delivered in 1939. During the following year most of the production problems were solved and the type was turned out in large numbers. As well as the Junkers plant at Bernburg, six plants from other companies built the aircraft under licence: the Heinkel plant at Oranienburg, the Arado plant at Brandenburg, the ATG plant at Leipzig, the Siebel plant at Halle, the Henschel plant at Berlin and the Dornier plant at Wismar. By their united efforts, the Luftwaffe received nearly 2,200 Ju 88s in 1940. Production of the Ju 88 reached its peak in 1943, when the Junkers, ATG, Siebel and Henschel plants turned out just over 3,300 of these aircraft. Total Ju 88 production ran to 14,676 aircraft, excluding prototypes.

Junkers Ju 88 production by year

1939	1940	1941	1942	1943	1944	1945	Total
100	2,184	2,619	3,094	3,301	3,013	355	14,666

ment comprised three MG 15 machine-guns firing from dorsal, ventral and nose positions. The variant had provision to carry an internal bomb load of 500 kg (1,102 lb).

Next came four further V-series prototypes, followed by the pre-production batch of A-0 series aircraft that began emerging from the production line at Bernberg in March 1939. These aircraft went to the service trials unit, Erprobungskommando 88, based at Rechlin near Berlin. The latter provided a nucleus of crews for the first operational Ju 88 unit, I. Gruppe of Kampfgeschwader 25, formed in August 1939.

At the outbreak of war on 1 September 1939, the Luftwaffe had taken delivery of about 15 pre-production Ju 88s. A few weeks later I./KG 25 was redesignated I. Gruppe of KG 30, and in the months to follow the first production Ju 88 variant, the A-1, entered service. The Ju 88A-1 had a top speed of 450 km/h (280 mph) at 18,000 ft (5486 m), and a range of 1,500 miles (2413 km). For short-range missions it could carry a maximum bomb load of 2400 kg (5,290 lb). The defensive armament comprised three 7.9-mm machine-guns, as fitted to the Ju 88 V5. In terms of its performance, its versatility in being able to make both dive- and horizontal-bombing attacks, and its general combat effectiveness, the Ju 88 was the finest medium bomber type in service with any air force at this time.

The first encounter

Between the last week in September and the first half of October 1939, I./KG 30 sent small forces of Ju 88s to attack Royal Navy warships operating in the North Sea. They achieved no success. Then, on the afternoon of 15 October, a German reconnaissance aircraft sighted a large warship – in fact the battle-cruiser HMS *Hood* – entering the Firth of Forth. The following morning, Hauptmann Helmut Pohl set out with a small force of Ju 88s (four or five aircraft) to attack the vessel. Pohl had strict orders that he could attack the

warship only if it was caught at sea. If the warship was in harbour it was on no account to be attacked, since this might endanger British civilians' lives.

The Ju 88s arrived over the Rosyth naval base to find the battle-cruiser in the act of entering harbour, and therefore off limits. Several smaller warships sat at anchor outside the harbour, however, and Pohl selected one of these for attack. He scored a direct hit with a 500-kg (1,102-lb) bomb on the cruiser *Southampton*. The weapon passed through three of the ship's decks at an oblique angle, before it emerged from the port side and entered the water. There it exploded, and the

This Ju 88 A-1 served with KG 1, which used the shield of Feldmarschall Hindenburg as its insignia. The A-1 was essentially similar to the A-0, but had three-bladed propellers instead of four-bladed units. Power was provided by the Jumo 211B-1.

Far left: Three of the four-man crew of a KG 76 Ju 88 crew pose for the camera. The crew compartment of the Ju 88 was cramped but provided for close crew co-operation. The pilot sat high up to port, with the bombardier low to starboard. The flight engineer faced aft behind the pilot, while the radio operator was low down on the starboard side.

Left: A common sight at Ju 88 airfields was this wheeled gantry that allowed the easy removal of the Jumo engine. Despite its appearances (thanks to the annular radiator), the Jumo 211 was an inline engine rather than a radial.

warship suffered minor damage. Leutnant Horst von Riesen attacked next, and scored a near miss on the destroyer *Mohawk*.

Then defending fighters, Spitfires of Nos 602 and 603 Squadrons, arrived on the scene. It was the first time, for either the Spitfire or the Ju 88, that these soon to be famous warplanes encountered an enemy aircraft in action. In short order the Spitfires shot down Pohl's aircraft and another Ju 88.

As von Riesen climbed away from the target, his radio operator reported a couple of fighters about a mile away, diving on them. There was no future in climbing, so the German pilot pushed down the nose of the bomber and attempted to outrun the fighters at low level, but it was to no avail. The Spitfires quickly caught up with the speeding Junkers and during the next several minutes the eight-gun fighters delivered attack after attack.

Low-level escape

The German crew was in no position to hit back on anything like equal terms. Only the radio operator's 7.9-mm machine-gun could be brought to bear on the attackers. The other two guns, pointing downwards and rearwards, and forwards, were useless. Jinking desperately to avoid the hail of bullets coming in his direction, von Riesen held the bomber just a few feet above the sea. It seemed only a matter of time before some vital part of the aircraft was hit. Moreover, the Spitfires were not the only hazard the Ju 88 crew now faced. Great fountains of water going up around the bomber told the

crew that anti-aircraft guns on the shore had joined the unequal struggle. The German pilot recalled:

"Now I thought I was finished. Guns were firing at me from all sides and the Spitfires behind me were taking turns at attacking. But I think my speed gave them a bit of a surprise – I was doing 400 km/h [250 mph], which must have been a bit faster than any other bomber they had practised against so low down – and of course I jinked from side to side to make things difficult for them. At one stage in the pursuit I remember I looked down and saw what looked like raindrops hitting the water. It was all very strange. Then I realised what it was: those splashes were marking the impact of the bullets aimed at me!"

Yet von Riesen and his crew had one valuable ally on their side: time. As the panting Jumo motors took the Ju 88 further out to sea, the pursuing Spitfires got progressively further away from their base. Then a well-aimed burst of machine-gun fire

Above: This Ju 88A-1 is loaded with four SC 250 bombs on its external racks. The window cut into the front of the starboard-side blister was the view-piece for the Lofte periscopic bombsight, which was used during level bombing attacks. For lining up dive-bombing attacks the pilot used the window in the port side of the cockpit floor. As described below, this window had a series of etched lines on it for range marking.

Right: Armourers prepare to load SC 250 bombs on to a Ju 88A-1 of KG 1. The right-hand bomb has been fitted with cardboard 'screamers' that created a howling sound during the bomb's fall to heighten the terror factor of the attack.

The dive attack

This section describes a typical diving attack manoeuvre employed by the Ju 88. Prior to the dive, the pilot levelled the aircraft precisely using the artificial horizon. For an accurate attack it was important to head into the wind during the dive. So, during their approach to the target, the crew would look out for smoke or other clues, to gauge accurately the direction of the wind.

The Ju 88 had a small window let into the cockpit floor, through which the pilot observed the target as the bomber moved into position to commence its dive. A series of parallel horizontal lines etched on the glass enabled him to judge his distance from the target. During his approach the pilot throttled back and set the propellers in coarse pitch, to prevent the engines overspeeding during the dive. Then he set the bomb release altitude, typically 1000 m (about 3,300 ft) on the contacting altimeter, and trimmed the aircraft for the dive.

As the target slid under the last horizontal line on the floor window, the pilot pulled the knob to extend the underwing dive brakes. The hydraulic system that operated the dive brakes simultaneously lowered and locked into position a tab on the underside of the elevator. That lifted the tail, and the bomber bunted into a 60° attack dive.

The set-piece manoeuvre was worked out so that, when the bomber was established in its dive, the target previously observed through the floor aiming window now appeared on the reflector sight in front of the pilot's eyes. The pilot then manoeuvred the aircraft to place the target in the centre of the illuminated aiming circle, and held it there during the dive.

If commenced from 3000 metres (just under 10,000 ft), the attack dive lasted some 15 seconds. When the contacting altimeter indicated 250 m (820 ft) above the bomb release altitude, a warning note sounded in the crew's earphones: prepare to release bombs. When the warning note ceased, the bomber was at the selected bomb release altitude. The pilot eased up the nose until the target lay under the base of the aiming circle, then pressed the bomb release button on his control column. As the last bomb fell away, a powerful spring returned the elevator trim tab to the neutral position. Now tail-heavy, the bomber pulled itself smoothly out of the dive and entered a slight climb. The pilot then retracted the airbrakes, selected fine pitch on the propellers, opened the throttles and climbed away.

This Ju 88A-4 served with the dive-bombing school at Tours in France, despite retaining the markings of its previous unit, 13./KG 54. The large numbers were for exercise purposes.

shattered the cooling system of the bomber's starboard engine. The latter belched a cloud of steam, forcing the pilot to shut it down and feather the propeller. The bomber's speed fell to about 180 km/h (110 mph), but fortunately for von Riesen and his crew the Spitfires were unable to exploit their advantage, having run short of fuel. They turned for home.

The German crew had shaken off the tormentors, but their situation was far from pleasant. Ahead lay a flight of nearly four hours to their base at Westerland, in an aircraft that lacked the power to climb on one engine. To keep the aircraft straight and level, von Riesen had to hold down the port wing and push against the starboard rudder pedal with all his strength, to prevent the bomber turning away from the good engine. To assist the pilot the navigator clambered into the nose, took off his belt, fixed it round the starboard rudder pedal and pulled for all he was worth.

As fuel was burnt, the Junkers gradually became lighter and von Riesen was able to coax the machine slowly higher. But even after three and a half hours, when a very relieved Ju 88 crew caught sight of the lights of Westerland, the bomber had not reached 2,000 ft (607 m). The almost exhausted pilot set the aircraft down on the airfield rather heavily, but otherwise the Ju 88 was little the worse for its experience. Fitted with a new radiator for the starboard engine, the aircraft resumed flying within a week.

Variations on the theme

In September 1938 the Ju 88 V7 took to the air, the proto-type of the Zerstörer (heavy fighter) variant of the aircraft. The main differences from the bomber version were a metal fairing in place of the glazed bomb aimer's nose, and the removal of the blister under the starboard side of the cockpit. The forward-firing armament comprised two 20-mm MG FF cannon and two 7.9-mm machine-guns. Instead of the bomber's crew of four, the fighter version carried three – pilot, flight engineer and radio operator. The flight engineer sat beside the pilot, his main task being that of cocking and reloading the drum-fed MG FF cannon. The Ju 88 V7 had a maximum speed of 500 km/h (311 mph) at 13,000 ft (3962 m),

or about the same as that of the fastest versions of the Messerschmitt Bf 110 then entering service. However, the range of the Ju 88 heavy fighter – 1,800 miles (2897 km) – was roughly three times that of its smaller counterpart. The modi-fied bomber was ordered into production as the Ju 88C.

During July and August 1939 a few early production Ju 88A-1 bombers were modified into heavy fighters, desig-nated as Ju 88C-0s. These aircraft had the solid nose of the V7, but retained the blister under the starboard side of the nose. The latter feature would remain with the Ju 88C throughout its life. A few Ju 88C-0s saw action during the Polish campaign, where they served in the long-range ground attack role.

The Ju 88C-1 was to have been fitted with BMW 801 engines, but at the time the latter were not available and the sub-type did not go into production. The next model to enter service, the C-2, had a revised armament of one 20-mm MG FF and three 7.9-mm guns firing forward, one 7.9-mm

Far left: Emblazoned with the death's head badge of KG 54, this Ju 88A-4 awaits a mission over England in its camouflaged dispersal at St André in France.

Left: Carrying an SC 1000 bomb and an SC 250, a Ju 88A-4 gets airborne. The port mainwheel has just begun its 90° rotation before retracting.

Below: Ju 88s climb away from their base towards England. If threatened the formation would tighten up to concentrate firepower.

Battle of Britain

This shot from the Battle of Britain, taken from another Ju 88, shows a Junkers under attack from a Hurricane. The fighter is just visible between the ring and bead sights of the MG 15 machine-gun.

Collecting the spoils of war – RAF personnel and a civilian pump fuel from the tank of a Ju 88 into a private car after the KG 54 bomber crash-landed near Tangmere in Sussex. Such practice was illegal, but a blind eye was usually turned by authority.

MG 15 firing rearwards and a bomb load of up to 500 kg (1,102 lb). The C-2 entered service in the early summer of 1940.

The excellent range and speed performance of the Ju 88 made it a useful addition to the strategic reconnaissance force, too. The reconnaissance variant, the Ju 88D, had the dive brakes removed, an additional fuel tank mounted in what had been the forward bomb bay, and two large reconnaissance cameras (typically an Rb 50/30 and an Rb 30/30) mounted in

what had been the rear bomb bay. Bomber units had first priority for the limited initial supplies of Ju 88s, however, and when the Battle of Britain opened only about a dozen of these aircraft were serving with front-line reconnaissance staffeln.

Battle of Britain

When the Battle of Britain opened in earnest in August 1940, the Luftwaffe possessed 12 bomber gruppen equipped with Ju 88s. Lack of space prevents a detailed account of Ju 88

RAF combat appraisal of the Junkers Ju 88

On the night of 23/24 July 1941, Ju 88A-6 4D+DH of I./KG 30 became thoroughly lost after attacking Birkenhead. The crew mistook the Bristol Channel for the English Channel and, running short of fuel, they landed at the next airfield they saw. It turned out to be the RAF airfield at Lulsgate Bottom near Weston-super-Mare. The pilot made a good wheels-down landing, thus presenting the RAF with its first fully intact Ju 88.

After its capture the Ju 88 was taken on RAF charge and given the serial number EE205. In the following August it went to the Air Fighting Development Unit (AFDU) at Duxford for combat testing against representative RAF types. EE205 went on to serve with No. 1426 (Enemy Aircraft) Flight at RAF Collyweston, until January 1945 when the Flight was disbanded. After a spell at the Central Flying Establishment at Tangmere, the Ju 88 went to No. 47 Maintenance Unit at Sealand for museum storage. Its fate thereafter is not known. Excerpts from the AFDU report are given below.

Brief description of the aircraft

The Junkers Ju 88 is a mid-wing twin-engined monoplane, fitted with Jumo 211 engines of 1,200 hp (895 kW) each. It is capable of carrying 6,400 lb (2903 kg) of bombs in which case its range is 550 air miles (885 km). Its maximum range is 2,200 miles (3540 km), but with a bomb load of only 1,100 lb (500 kg) it is fitted with dive brakes, which permit angles of dive of up to 70°. There is a fully automatic pilot and also provision for a second control column, which the bomb-aimer can use. Other equipment includes de-icing for airscrews and tailplane, and provision for balloon cable-cutters in the wings.

Cabin: The crew of four are positioned together in a small compartment well forward. The pilot and the upper gunner, who is also the wireless operator, have comfortable seats and sufficient room to carry out their duties; but conditions for the bomb-aimer, and the lower gunner who lies in a gondola slung beneath the starboard side of the cabin, are cramped and uncomfortable. It is not thought possible for the lower gunner to remain at his gun for more than an hour at a time, and he has a small rest seat just above his position. The bomb-aimer sits in a cramped position beside the pilot when he is not lying over his sight, which is uncomfortably close to the rudder pedals and bomb switches.

Armament characteristics

The guns on the aircraft available for trials are four MG 15s [7.9-mm] in gimbal mountings. Two of these are in the upper rear part of the cabin, set in separate rotatable bullet-proof window rings. The third gun is in the gondola firing aft, and the fourth fires forward mounted on the starboard side of the front of the cabin. The number of rounds available is 1,650 in 22 magazines of 75 rounds each. The magazines are fitted on racks in the side and back of the cabin and in all spaces where there is no other equipment. The armament appears to be the latest and most efficient the Germans have adopted for this aircraft.

Sights: The sights for all guns are 60 mph [deflection] ring and bead. The bead is nearest the gunner's eye and the ring at the muzzle end of the gun.

Functioning: One gun was specially mounted and fired on the ground at stop butts. The three rear guns were fired from the aircraft over the sea. No stoppages occurred and the recoil was negligible at any angle. The rate of fire is 1,000 rounds per minute and there is little smoke or fumes. The ammunition did not appear to be loaded in any definite order, but approximately one round in five was red tracer or green incendiary tracer. The remainder consists of ball, armour-piercing and high explosive. A night firing test was carried out and the tracer did not have a particularly dazzling effect. The general impression is that the gun is very simple and an extremely effective weapon. The time taken to change magazines is 5 seconds.

Area of fire: The upper rear guns cover a good field of up to about 75° on each beam and 35° in elevation. The lower gun is capable of being fired through an angle of 40° each side from dead astern – but firing to starboard is easier than to port, owing to the position of the gunner with the stock to his right shoulder. The maximum depression of 30° is only obtained astern and is reduced to 15° on each side. There is no blind spot behind the tail. The front gun is very restricted and can be locked if necessary in a horizontal position, presumably so that the pilot can fire it, though in this aircraft there was no method for him to do so.

Armour: The crew is extremely well protected against attacks from the rear and fine quarter. The whole of the back of the cabin is protected by armour and bullet-proof glass as far down as the upper gunner's knees. The lower gunner has a dome of armour plate below him and a semi-circular piece of armour which rotates with his gun, to protect his face. The pilot has an armour plate headpiece to his seat, and the back is protected from the rear and fine quarter. The underneath of the upper gunner's seat is armoured. As the crew are well forward, they are also protected by the self-sealing fuselage fuel tank and wireless gear aft of the cabin, which would probably stop bullets from dead astern. All fuel and oil tanks are self-sealing, but there is no protection for the engines from astern or ahead.

Tactical trials

Flying characteristics: The Ju 88 is remarkably manoeuvrable for an aircraft of its size and wing loading. The controls are light and positive and all are assisted, with the result that they do not become heavier at high speeds but allow the aircraft to take quite violent evasion even at the end of a long dive. After gaining experience a pilot should be able to fight the aircraft quite well and give difficult deflection shooting to opposing fighters. The Ju 88 is capable of a high top speed. In comparative trials with a Blenheim IV in which both aircraft were unladen, it appeared about 25 mph (40 km/h) faster both at 2,000 and 15,000 ft (607 and 4572 m). It accelerates quickly in a dive. The dive brakes are easily applied and are very effective. In dives of 6,000 ft (1829 m) at an angle of 60° the maximum indicated airspeed was never more than 265 mph (426 km/h), and there was full control for aligning the sights. The pull-out was automatic.

The aircraft trims easily to fly hands off and instrument flying is comfortable. No night flying was carried out during the trials, but the cockpit lighting appears satisfactory by night, without reflections on the perspex. Single-engined flying was carried out with each airscrew feathered in turn. The Ju 88 is not comfortable to fly on one engine but will maintain height when unladen at about 160 mph (257 km/h) indicated. It does not turn easily against the live engine and accurate recovery from a turn with it is difficult. It is thought, therefore, that aircraft which managed to get away after combat with only one engine intact are unlikely to be landed back in their own territory safely except by very experienced pilots.

View: The pilot's view is not good by our standards. He has to sit well up with his head almost touching the roof and even then he can see only forwards and to the port side (on which he sits) with any comfort. His view to starboard is obstructed on the bow by the front free gun and by cabin former ribs to the rear. He sits aft of the airscrews but can see over the engines more easily than in a Blenheim. The best point about his vision is that the windscreen is made of a number of flat direct vision panels, none of which cause distortion. The upper gunner's all-round view is good. He is of course blind beneath the wing but the lower gunner can cover the view downward to the rear adequately. The only slightly blind spot, if the bomb-aimer is not lying in position, is forward and below where sights and fittings mar the view.

Formation flying: The Ju 88 is easy to fly in close formation, being responsive to the throttle and decelerating quite well. It is necessary, however, to be on a level with or below the leader as the engines interrupt the downward view to the sides.

Fighting manoeuvres

General: The Ju 88 was flown in combat with Hurricane II and Spitfire I aircraft

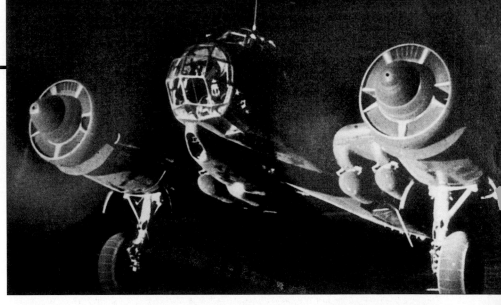

operations during the battle, but four of its early actions deserve mention.

The first large-scale Ju 88 attack on a target in England took place on the morning of 11 August, when about 50 bombers from I. and II. Gruppen of KG 54 set out to attack the Royal Navy base at Portland. The large German fighter escort held off the defending Spitfires and Hurricanes for a time, allowing the Ju 88s to dive-bomb the facility. The raiders wrecked two large oil storage tanks, and caused other damage. Then defending fighters got through to the bombers and shot down five Ju 88s in rapid succession. Among those lost was the Kommandeur of II./KG 54, Major Kurt Leonhardy.

Portsmouth attack

On the following day the Ju 88s fared even worse. All three gruppen of KG 51 took off to attack Portsmouth naval dockyard and the radar station at Ventnor, again with a large escort of Bf 109s and Bf 110s. The Ju 88s delivered an accurate attack on the dockyard causing severe damage, as did that on the radar station. Again there was a large-scale reaction by defending fighters; they and AA gunners shot down 10 Ju 88s and inflicted damage on three more. Among the senior personnel lost was Oberst Johann-Volkmar Fisser, the Geschwader Kommodore.

Three days later, on 15 August, Ju 88 units suffered another heavy drubbing. At mid-day about 50 Ju 88s of all three gruppen of Kampfgeschwader 30 set out from Aalborg in Denmark to attack the RAF bomber airfield at Driffield in Yorkshire. The distance involved placed the attack beyond the radius of action of the Bf 109, but the Luftwaffe High Command was confident that an attack on a target in northern England would take the defenders by surprise. In case a few

For more than six months from September 1940, the Ju 88 units were engaged in nocturnal attacks against British cities in what became known as the Night Blitz. Here a Ju 88 taxis for a mission with four SC 250s.

belonging to this unit. Camera guns were used by the fighters. The Ju 88 was flown without bomb loads at all times, the total weight being 22,500 lb (10206 kg).

It was found that in no type of attack can the fighter be met by more than one gun at a time, and since the gun is hand-held on a gimbal mounting continuous accurate fire is very difficult. The tail-down flying attitude of the Ju 88 should always be borne in mind by fighter pilots. In level flight the tail appears to be down about 10° and therefore in deflection shooting careful allowances must be made. This tail-down attitude also allows the upper gunner good shots to the rear during an astern attack.

Astern attacks: It was found impossible for the attacking fighter to hide behind the tailplane or rudder. The upper gunner could change guns more rapidly than the fighter could move from side to side, and the lower gunner could fire at all times when the fighter went below the upper gunner's field of fire. If the attack is made on the level, the upper gunner has an easy shot due to the tail-down attitude of the Ju 88 in flight. Attacks should therefore be made just below the level of the tailplane. Since the rear of the cabin is so well armoured astern, attacks with 0.303-in (7.7-mm) guns should be directed at the engines which are unprotected.

The evasive manoeuvres attempted by the Ju 88 included skidding, undulating, high-speed dives and turns. Of these skidding, though easy to carry out due to the light rudder, was not particularly effective and was usually detected by the fighters, but undulating caused the fighters serious sighting difficulties. The undulation was easy to effect as the Ju 88 does not lose its engines when the control column is pushed forward causing negative g. The fighters found it best to remain on the top level of the undulation so as to fire at the Ju 88 at its no-deflection point and during the first part of the downward dive. The Spitfire's engine kept cutting when it tried to follow the Ju 88.

Diving appears from analysis of recent combat reports to be the favourite method of evasion adopted by the Ju 88. The Ju 88 therefore was climbed to 15,000 ft (4572 m) with a Spitfire and a Hurricane about a mile behind and one Hurricane as escort in formation. It then dived losing about 1,500 ft (457 m) per minute at full power and the Spitfire closed up to firing range after a loss of 8,000 ft (2438 m), having covered 20 miles (32 km). The attacking Hurricane, however, had closed the range only slightly. During the dive the escorting Hurricane never exceeded 4½ pounds boost pressure or 2,550 rpm. The maximum speed of the Ju 88 was 305 mph (491 km/h) indicated.

Rudder control on the Ju 88 was still very light and the tail could be swished gently to give the rear gunner a good view.

The Ju 88 also evades by means of turns, especially to gain cloud cover. It was found that it could turn quite sharply even at high speeds but the fighters never had any trouble in keeping their sights on it. However, the controls of the Ju 88 are so good that the steepness of the turns can be varied easily and sideslip added so that difficult shots are provided for the fighters.

Quarter and Beam attacks: The Ju 88 has a little armour to protect the cabin and occupants from fine quarter attacks, so that the fighter armed with 0.303-in guns only must concentrate on obtaining hits during the quarter attack between 45° and 30°. The upper gunner can bring his gun to bear quite easily almost on to the full beam, but the lower gunner's field of fire is only 40° on each side. In addition, the lower gunner can see and fire more easily to starboard than to port and therefore attacks should be made if possible

on the port quarter and from slightly below the Ju 88, and the breakaway should be downwards on the port side. A further advantage in attacking from the port is that the pilot sits on that side and is fairly exposed to quarter attacks.

Head-on attacks: The Ju 88 is entirely unprotected from head-on attacks and with its large circular radiators presents a good target. This type of attack is ideal if the fighter can get into position unobserved, but will be impossible if he has been seen beforehand, owing to the Ju 88's high speed. The field of fire of the front gun of the Ju 88 is very limited and it is difficult to use, owing to the cramped position of the bomb aimer. His close proximity to the pilot makes firing to port easier than to starboard.

Ju 88 as a dive-bomber: The Ju 88 was dived at angles between 50° and 60° with the assistance of the dive brakes. Attacks were made by fighters from which it was found possible to fire an effective burst during the dive, before overshooting, but this prevented the fighter getting a more certain shot as the Ju 88 pulled out of the dive. It would seem best, therefore, for the fighter to spiral down during the dive and engage the Ju 88 as it pulls away. The Ju 88 also makes use of its dive brakes as a very effective evasive manoeuvre, especially when it desires to attain cloud cover.

Slipstream: The slipstream is not strong at any point. Fighters were able to come in to less than 100 yards (91 m) astern and maintain quite steady bursts.

Low flying: The Ju 88 is fully controllable above about 160 mph (257 km/h) and can therefore be used for low-level attacks.

Night fighter attacks: Our fighters should normally attack from astern and below as the lower gunner's vision and arc of fire are restricted. For free-gun fighters [i.e. Defiants] there is a large blind spot below the centre section that can safely be approached from the beam.

EE205 is seen in full RAF markings during trials with the bomber. The aircraft was built as a Ju 88A-6 with provision for balloon cable-cutting equipment (note the mountings on the nose, although the equipment was only rarely used), but was otherwise similar to the Ju 88A-5.

Above: Ju 88s headed for Britain by day and by night. The Luftwaffe turned primarily to night attacks in the autumn of 1940 as the bombers had suffered considerably at the hands of RAF day fighters. On the other hand, Britain's nocturnal defences were much weaker.

Right: For night operations many bombers were given a crudely applied coat of black distemper to render them less visible. This Ju 88A served with KG 1.

Right: The Ju 88C had a solid nose housing three MG 17 7.9-mm machine-guns and a single MG FF/M 20-mm cannon. Two additional cannon, each with 120 rounds, were mounted in the gondola, as seen on these Ju 88C-6s. All guns were depressed 5°.

Below: These aircraft are from NJG 2, based in Holland, from where they flew on night intruder missions over the UK.

defending fighters took to the air, the force included about a dozen Ju 88C heavy fighters which were on the strength of the Geschwader.

Thanks to radar, the defenders were not taken by surprise and one squadron of Spitfires and half a squadron of Hurricanes intercepted the raiders near the coast. It would seem that the Ju 88Cs attempted to block the RAF fighter attacks, for they suffered disproportionately heavy losses. In the ensuing action five of the heavy fighters were shot down, one was wrecked when it crash-landed in Holland and one more returned with battle damage. The raiders delivered a concentrated attack on Driffield, which damaged four hangars and destroyed 10 Whitley bombers on the ground. The bomber force lost two Ju 88A-5s, and one more returned with battle damage.

That afternoon Lehrgeschwader 1 sent about 60 Ju 88s from I.and II. Gruppen, with an escort of Bf 110s, to attack the airfields at Middle Wallop and Worthy Down. Three squadrons of Hurricanes and one of Spitfires engaged the raiders and shot down eight of them. The attackers caused little damage at either airfield. Some bombers missed their target and instead bombed Odiham airfield, afterwards reporting they had hit the airfield at Andover.

The lesson of these actions was that while the Heinkel He 111 and the Dornier Do 17 were slower and less manoeuvrable than the Ju 88, they were better able to withstand battle damage.

The Night Blitz

Following the switch from airfield targets to London itself, on 7 September 1940, the capital came under attack both by night and by day. No Ju 88 bomber units took part in the great daylight action on 15 September. However, these units were active over the capital on the night of 15th/16th. Seven aircraft of Kampfgruppe 806 attacked London from 8 pm, followed by 10 from III./LG 1 at 8.50 pm, 10 from I./LG 1 at 9.46 pm, 10 more from I./KG 54 at 11.12 pm and eight from II./KG 51 at 11.50 pm.

The main targets that night were the dock areas to the east of the capital. At this time Britain's night air defence system was weak. To exploit this, and to create maximum possible disruption on the ground, bomber crews were briefed to remain over the target area as long as they reasonably could, releasing one bomb every five minutes or so.

That night RAF Fighter Command flew 64 interception sorties, but neither the fighters nor the AA gunners claimed a kill. However, two Ju 88s of II./KG 51 failed to return from the raid, the fate of these aircraft and their crews is not known.

The night attacks on targets in Great Britain continued until May 1941, when the Luftwaffe bomber force moved eastwards to prepare for its next major campaign.

Night intruder missions

In May 1940, RAF Bomber Command commenced night attacks on industrial targets in Germany. Göring ordered Oberst Kammhuber to form a night fighter Geschwader, and at the end of July Nachtjagdgeschwader 1 comprised a gruppe of Bf 110s, one of Bf 109s and II. Gruppe with Ju 88Cs.

In September 1940 II./NJG 1 was re-designated I./NJG 2, and the unit's Ju 88Cs began night intruder operations against Bomber Command bases in England. Aircraft on the final approach to landing had little reserve speed for manoeuvrability, and were virtually 'sitting ducks' – a fact known only too well to the crews. The intruders shot down few bombers, but they caused much disruption. Many aircraft were damaged in heavy landings, for there could be no thought of going round again if intruders were thought to be about, no matter how bad the approach.

'Big Bang' at Piraeus

Following its attacks on Great Britain in the autumn and winter of 1940, I./KG 30 took its Ju 88s to Gerbini in Sicily where it became part of the Fliegerkorps X operating in support of Italian forces in Albania and Greece. On 6 April 1941 Hitler launched Operation Marita, the invasion of Greece. Late that afternoon I./KG 30 launched about 20 Ju 88s to drop mines in the approaches to Pireaus, the port of Athens, and attack ships in the harbour waiting to unload.

That night the weather over western Greece was atrocious. Dense storm clouds towered high above the bombers, causing severe turbulence and icing. From the bombers' radio aerials and wing tips there danced the eerie fluorescent flame tips of St Elmo's fire. Some bombers encountered severe airframe icing, and had to jettison their loads and return to Gerbini.

Once they were past Corinth, the remaining bomber crews found clear skies. In the light of the half moon Hauptmann Hans-Joachim Herrmann had no difficulty in finding the port of Piraeus, and he planted his mines in the main navigation channel. Then he climbed away and circled clear of the port, to enable his observer Leutnant Heinrich Schmetz to measure the direction and speed of the wind and set it on his Lofte bombsight. Herrmann climbed the Junkers to 10,000 ft (3048 m). He lined himself up on the port, cut the throttles, and ran in for a silent gliding attack. Schmetz had picked out a large freighter moored against the jetty. Glued to the eyepiece of his bombsight, he called out precise course corrections over

Above: With Mount Etna in the background, a Ju 88A-5 of III./KG 30 is prepared at its base. The gruppe was dispatched from Norway to Sicily to assist with the operations in the Greek campaign, before returning north for the big Arctic Convoy battles.

Left: Hauptmann Hajo Herrmann of I./KG 1 – whose account of the Piraeus raid is recorded opposite – is seen in his Ju 88 with navigator Leutnant Hienrich Schmetz while stationed at Gerbini in Sicily. The view shows good detail of the flexible mount for the forward-firing MG 81 machine-gun, and KG 30's diving eagle badge.

the intercom. At 3,000 ft (914 m) Schmetz released his bombs in a stick, then Herrmann banked the aircraft to observe the results.

Suddenly the ship disappeared in a huge sheet of flame, and the sky over the port lit up as if in broad daylight. Seconds later

---

Junkers Ju 88s in service with operational units, 24 June 1941

This section gives the strength of Junkers Ju 88 front-line units on 24 June 1941, three days after the start of the attack on the Soviet Union. On that day the front-line units reported a total of 945 Ju 88s (659 serviceable) on strength. A majority, 707 aircraft, served with bomber units and the type made up the backbone of the twin-engined bomber force. A further 174 Ju 88s served with reconnaissance units, and the type made up the backbone of the long-range reconnaissance force also.

The first figure gives total number of aircraft, the second figure gives the number serviceable

BOMBER UNITS

Lehrgeschwader 1		
Stab	1	1
I. Gruppe	35	4
II. Gruppe	25	11
III. Gruppe	27	11
Kampfgeschwader 1		
II. Gruppe	29	27
III. Gruppe	30	29
Kampfgeschwader 3		
Stab	1	1[1]
I. Gruppe	38	32
II. Gruppe	38	32
Kampfgeschwader 30		
I. Gruppe	34	19
II. Gruppe	31	22
III. Gruppe	15	6
Kampfgeschwader 51		
Stab	2	2
I. Gruppe	22	22
II. Gruppe	36	29
III. Gruppe	32	28
Kampfgeschwader 54		
Stab	1	1
I. Gruppe	34	31
II. Gruppe	36	33
Kampfgeschwader 76		
I. Gruppe	31	22
II. Gruppe	30	25
III. Gruppe	29	22
Kampfgeschwader 77		
Stab	1	1
I. Gruppe	30	23
II. Gruppe	31	23
III. Gruppe	29	20
Kampfgruppe 606	29	13
Kampfgruppe 806	30	18

NAVAL CO-OPERATION UNITS

Küstenfliegergruppe 106	17	4[1]
Küstenfliegergruppe 506	11	4

RECONNAISSANCE UNITS

Aufklärungsgruppe 10		
3. Staffel	6	5[1]
Aufklärungsgruppe 11		
3. Staffel	6	5[1]
4. Staffel	9	8
Aufklärungsgruppe 14		
4. Staffel	9	6
Aufklärungsgruppe 22		
1. Staffel	7	5
2. Staffel	9	8
3. Staffel	9	6
Aufklärungsgruppe 33		
1. Staffel	9	7
3. Staffel	9	9
Aufklärungsgruppe 120		
1. Staffel	12	6
Aufklärungsgruppe 121		
1. Staffel	12	5
Auflärungsgruppe 122		
1. Staffel	7	7[1]
2. Staffel	9	7[1]
3. Staffel	8	3[1]
4. Staffel	8	8[1]
5. Staffel	9	8
Aufklärungsgruppe 123		
1. Staffel	9	6[1]
2. Staffel	6	2[1]
3. Staffel	4	4[1]
4. Staffel	8	5[1]
Aufklärungsgruppe 124		
1. Staffel	9	4[1]

NIGHT FIGHTER UNITS

Nachtjagdgeschwader 2		
Stab	4	4
I. Gruppe	32	15[1]

Reporting to Luftwaffe High Command

Aufklärungsgruppe der Oberkommando der Luftwaffe[2]	91	57

(Ju 86P, Ju 88, He 111, Bf 109, Bf 110, BV 142, Do 17, Do 215, Do 217)

1. Unit operated other type or types.
2. This large reconnaissance unit operated under the direct control of the Luftwaffe high command, and at this time it served mainly on the Eastern front.

Ju 88s in the Mediterranean theatre

Above: Bomb-laden Ju 88s of KG 54 head for Benghazi in Libya in the autumn of 1942 from their base at Catania, Sicily. The geschwader was also involved in the attack on the Soviet Union in 1941.

Below: From their bases in Sicily the Luftwaffe's Ju 88 force ranged far out over the Mediterranean, attacking both ground and ship targets. Targets in Malta and North Africa were regularly attacked.

I./Lehrgeschwader 1 was based at Malames in Crete when this pair of the unit's aircraft was photographed. The national insignia were hastily overpainted to reduce conspicuity in the night bombing role.

Left: Lehrgeschwader 1 ('L1' codes) bore the brunt of operations in the Mediterranean, serving in the theatre between early 1941 and mid-1944. The unit initially specialised in anti-shipping attacks, but later broadened its duties to include attacks on Commonwealth forces in North Africa. Returning to western Europe to oppose the Normandy landings, LG 1 was the only major Ju 88 unit not to see service in Russia.

the shock wave from the explosion buffeted the aircraft. Herrmann commented, "It sounded as though there was a man with a big hammer outside, and beating it against our aircraft." In the harbour there were further explosions, with balls of fire hurtling out in all directions.

The ship was the 12,000-ton freighter *Clan Frazer* which had arrived earlier in the day carrying, among other things, 350 tons of assorted munitions. About a hundred tons of the explosives had been offloaded when work ceased at dusk. The 250 tons of munitions remaining aboard the vessel had just detonated, with a force powerful enough to smash windows in Athens seven miles away. *Clan Frazer* disintegrated, and 10 other ships totalling 41,000 tons were also wrecked.

Herrmann attributed his survival to the curiosity that made him bank the aircraft after bomb release, with the result that the Junkers avoided much of the force of the explosion. The German pilot flew an orbit over the harbour, trying to discover what had happened. But, apart from some scattered fires, everything below seemed still. Stunned by the shock of the explosion, the defending gunners had ceased firing and the port's searchlight pointed in all directions, their beams motionless.

Then a solitary gun opened fire - and shot out Herrmann's port motor. "So justice is meted out," he commented. Flying at 1,000 ft (305 m) with an aircraft that would not climb, there could be no question of attempting to regain his base at

Gerbini 500 miles (805 km) away to the west. The only chance of survival was to divert to the island of Rhodes, 220 miles (354 km) to the east, which was in Italian hands. With some difficulty Herrmann reached the island on his remaining engine, and made a successful landing.

The explosion of the freighter *Clan Frazer*, with a force of almost nuclear dimensions, was to have serious consequences for the British forces in Greece. It flattened much of the port of Piraeus, and greatly reduced its capacity to offload cargoes. In the words of Admiral Andrew Cunningham, commander of the Royal Navy's Mediterranean Fleet, it was "a shattering blow". By depriving the British forces of the only reasonably equipped port through which supplies could be landed in Greece, it materially shortened the German campaign to overrun the country.

The Eastern front

During May 1941 the raids on Britain tapered off as the Luftwaffe bomber force repositioned the main part of its strength in readiness for Adolf Hitler's next great adventure: the invasion of the Soviet Union. The onslaught opened during the early morning darkness of 22 June and achieved complete surprise. Eighteen of the 23 bomber gruppen equipped with the Ju 88 were committed to the enterprise.

Before dawn small forces of Heinkel He 111s, Dornier Do 17s and Junkers Ju 88s, flown by picked crews, attacked the more important Soviet airfields. Their aim was to disrupt activity at the airfields until larger attack forces arrived after daybreak. Even so, the majority of Soviet air units had not received the warning order when the main attack opened. As a result, the raiders found hundreds of Soviet combat aircraft sitting on the ground in neat rows.

Above: An early Ju 88 churns up the dust somewhere in the Med. It has the original style of rear-facing armament.

Above left: This Ju 88A-4 of KG 54 was seen at Bergamo in Italy in 1943. It features the unusual Wellenmüster (wave mirror) camouflage.

Left: This Ju 88A-4 of KG 54 at Catania carries the Kleiderbombe capsule under the wing. This was used to carry the crew's clothes during ferry flights.

Below: Some Ju 88A-14s, like this 1./KG 77 example, were fitted with 20-mm MG FF cannon in place of the bombsight for anti-ship attacks. It operated from Sicily in late 1942.

Above: A Ju 88A of 8./KG 77 cruises over Russia in the summer of 1941, shortly after the launch of Operation Barbarossa. Despite the overall military successes achieved in the first few weeks of the campaign, the fighting took a toll on operational strength. KG 77 fought for a year in the East before being transferred to Sicily.

Right: On 16 August 1941 this Ju 88 of I./KG 51 suffered a ramming attack by a Soviet fighter which tore off the starboard tailplane and badly twisted the rear fuselage. In a testament to the flying skills of pilot Leutnant Unrau and the sturdiness of the Ju 88's airframe, the aircraft was nursed back to friendly territory where the crew bailed out safely.

The target for the Ju 88s of III./KG 76 was the airfield at Krudziai in Lithuania. Leutnant Dieter Lukesch remembered:

"The skies were beautifully clear, with visibility almost unlimited. Soon after we took off we could see the front line quite clearly, marked by fires and the smoke from bursting shells. Once we had passed the front, however, there was no Flak. We did not have, nor did we expect to need, an escort for the first attack. As we passed other airfields we saw Russian fighters taking off, but they climbed somewhat slower than our cruising speed so we soon left them behind."

The Ju 88s cruised at 10,000 ft (3048 m), each carrying four 250-kg (550-lb) and 10 50-kg (110-lb) general-purpose bombs. When they reached the target the bombers moved into line astern and attacked in shallow dives. Lukesch saw more than a score of Tupolev SB-2 bombers drawn up in a line along one side of the airfield:

"There was no flak, and even though the war had been in progress for about three hours it seems that we had achieved surprise. As we approached for the first attack we could see ground crewman standing on the wings refuelling the aircraft, looking up in curiosity as we ran in. As the bombs started to explode they made a hasty retreat into the nearby forest."

Lukesch was about to release his bombs when he saw another Junkers converging on him from the right.

"I had to break away, make a circuit and attack at the end of the force. I ran in as the last aircraft in the gruppe to attack. By then several aircraft on the ground were burning and there was quite a lot of smoke, but the line of trees behind the aircraft helped me to line up on some planes that had not been hit before. During the attack my observer fired at the enemy planes with his machine-gun.

"As we pulled away after the attack some Russian fighters appeared on the scene, Ratas and Gulls [Polikarpov I-16s and I-15s]. Although they got close they did not fire at us, perhaps they did not have any ammunition. With my greater speed I soon left them behind."

Right: The sight greeting many Luftwaffe bomber crews in the opening attacks of Barbarossa were neat rows of Soviet aircraft on their fields – ripe for attack. This is a row of 23 Tupolev SB-2 light bombers drawn up at Krudziai in Lithuania and photographed on 21 July 1941, the day before the attack by III./KG 76. The events of this raid are recounted on this page by Leutnant Dieter Lukesch.

Far right: Dieter Lukesch (right) is seen later in the war, still flying Ju 88s but having attained the rank of Hauptmann.

The operations on the new battlefront would be characterised by high sortie rates that, maintained over a long period, led to heavy cumulative losses. During the first 100 days of the campaign the Luftwaffe lost, on average, 16 aircraft of all types each day. In the short term that was a sustainable loss rate, but extended over a long period it bled the force white. In the first hundred days the Luftwaffe lost around 1,600 aircraft – a figure that exceeded the German production of combat aircraft over that period.

Ground attack

Such was the pace of the fighting that twin-engined bombers were often sent to operate in the close support role, to assist German army units. These relatively large aircraft, flying at low altitude, were vulnerable to ground fire and suffered serious cumulative losses. Each Ju 88 lost in this way represented a double loss compared with a small ground attack aircraft or a dive-bomber. Not only was it much more expensive, but it carried a crew of four trained men that would need to be replaced.

The use of twin-engined bombers in this way was justified in exceptional circumstances, for example to disrupt an imminent enemy attack that might overrun a friendly position. Yet, as the war progressed, medium bombers were sent on close support missions on many occasions when no critical situation existed. Luftflotte commanders feared that if they failed to commit their forces in action often enough, they would lose units to other parts of the front. Thus, even in quiet sectors, there was pressure to keep air units continually in action even when there was little chance of achieving a decisive result. Air

Recce 88s in Russia

crews were not allowed adequate time for rest or refresher training, though the latter was particularly needed considering the uneven training replacement crews were getting during the mid-war period.

Top: Ju 88D reconnaissance aircraft of 2./AufKlGr. 22 rest on a snowy Russian airfield. The unit was active on the southern sector of the front for most of the Russian campaign. Note the protective covers for the engines.

Above: Ground crew dig a Ju 88D out of the snow. Visible are the three camera windows under the belly, and the long-range drop tanks regularly carried on recce missions.

Left: Ju 88s (and their bombs) share an airfield in Russia with Ju 52/3m transports.

Ju 88 'specials'

Ju 88 V7 high-speed transport

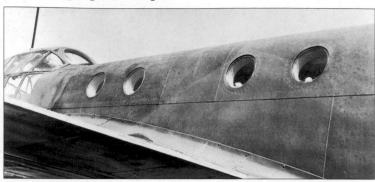

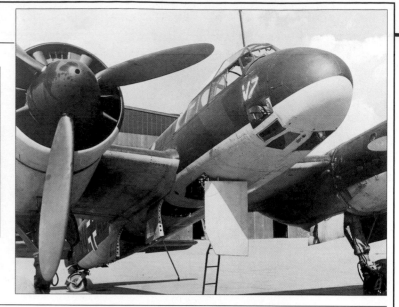

Having served as a prototype for the Ju 88C fighter, with four guns mounted in the starboard side firing forwards through a glazed nose, the V7 was converted to serve as a high-speed passenger transport. It had cabin windows cut into the fuselage (above) and was fitted with a streamlined 'solid' nose (right).

Engine testbeds

With its tall undercarriage providing good ground clearance, and its external bomb racks providing a ready-made mount, the Ju 88 was a natural choice to test the German jet designs. The poor-quality photo below shows Junkers's own Jumo 004 on a Ju 88, while at right is the rival BMW 003 under a Ju 88A-5.

'Dobbas' freight-container

The 'Dobbas' was devised as a method of rapidly turning the Ju 88 into a freight carrier. A rack could be mounted between the two inner ETC bomb racks (above), which in turn supported a collapsible container (right), seen here with an artillery gun inside.

Rocket-assisted take-off

The Ju 88A-2 was fitted with the ability to carry Walter HWK 500 Rauchgeräte rocket packs (above) under the outer wings to assist overload take-offs. The packs were jettisoned after use and descended by parachute (right).

Junkers Ju 88A-5

9. Staffel/Kampfgeschwader 30
Gilze-Rijen, 1941

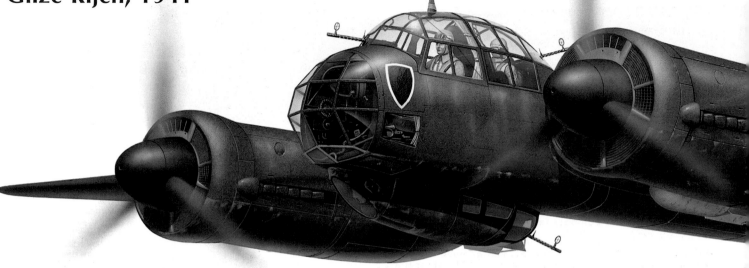

Gun armament
The Ju 88's standard defensive armament comprised a 7.9-mm machine-gun mounted in the starboard windscreen fired by the pilot, and similar aft-firing guns operated by the flight engineer (upper) and radio-operator (ventral gondola). The initial weapon was the MG 15 but this soon gave way to the much better MG 81. Results from combat saw additional guns being added, such as lateral-firing MG 15s in the cockpit glazing, and a single MG 131 or two MG 81s firing forwards from the noze glazing and operated by the bombardier. The upper rear armament was changed in mid-1941 to two MG 81s in armoured circular mounts. A few Ju 88A-4 and A-14s operating in the anti-shipping role were fitted with a fixed, forward-firing MG FF 20-mm cannon to suppress fire from the ship and to attack small vessels.

Ju 88 variants

Ju 88 prototypes: The first prototype, the **V1**, made its first flight on 21 December 1936. It was powered by two Daimler Benz DB 600A engines rated at 1,000 hp (746 kW) for take-off. The second prototype, the **V2**, was essentially similar. The **V3** was powered by Junkers Jumo 211A engines also rated at 1,000 hp for take-off. This was the first Ju 88 to carry armament, and it had the rear cabin raised to permit the installation of a rearward firing 7.9-mm machine-gun. The **V4** was the first variant with the 'beetle's eye' flat optical panels in the nose. The **V5** was a specially modified record breaking aircraft, with a streamlined airframe and powered by Jumo 211B-1 engines. The **V6**, **V8**, **V9** and **V10** were production prototypes to test features of the Ju 88A series of bombers. The **V7** was the prototype for the Ju 88C heavy fighter.

Ju 88A-0: pre-production series of 10 bomber variants powered by Jumo 211B engines. Deliveries commenced in March 1939.

Ju 88A-1: the first production bomber variant, similar to the A-0.

Ju 88A-2: similar to the A-1, but with provision to carry booster rockets under the wings to allow it to take off carrying heavier loads.

Ju 88A-3: conversion trainer version of the A-1, fitted with dual controls.

Ju 88A-4: first major rework of the Ju 88, it was produced in large numbers. Main production version powered by the Jumo 211J engines rated at 1,350 hp (1007 kW) for take-off. The wing span was extended by 1.62 m (5 ft 4 in) to improve handling, and the undercarriage was strengthened to permit the carriage of heavier loads.

Visible on the nose of this Ju 88A-6 are the cable-cutter mounting posts.

Ju 88A-6/U: conversion for the maritime role with Jumo 211J engines and FuG 200 Hohentwiel radar.

Ju 88A-7: conversion trainer version of the A-5, fitted with dual controls.

Ju 88A-8: modification of the A-4, with *Kuto-Nase* balloon cable-cutters fitted to the wings and the crew reduced to three. Built in small numbers.

Ju 88A-9: tropicalised version of the A-1, fitted with desert survival equipment, sand filters and sun blinds.

Ju 88A-10: tropicalised version of the A-4, fitted with desert survival equipment, sand filters and sun blinds.

Ju 88A-11: similar to A-10, but with further changes to suit it for operations in hot areas.

Ju 88A-12: A-4 modified as a conversion trainer with dual controls. The ventral cupola, the dive-brakes and the armament were removed.

Ju 88A-13: A-4 modified for the ground attack role. This variant had the dive-brakes removed and it carried additional armour. There was provision to carry up to 16 forward-firing 7.9-mm machine-guns in pods mounted under the wings.

Ju 88A-14: similar to the A-13, but with yet more armour protection for the crew. Some of these aircraft fitted with a fixed forward-firing 20-mm MG FF cannon in the ventral gondola.

Ju 88A-15: similar to the A-4, but with an enlarged bomb bay and the ventral gondola removed.

Ju 88A-4 of III. Gruppe/Lehrgeschwader 1. The A-4 was the most important bomber version, with Jumo 211J engines and extended wings.

Ju 88A-5: interim variant with the airframe modifications of the A-4. Due to production difficulties with the Jumo 211J engine, this variant was powered by Jumo 211B, G or H engines, all rated at 1,200 hp (895 kW) for take-off.

Ju 88A-6: similar to the A-5 but fitted with a fender in front of the nose and wings, with cutter at each end to slice through balloon cables. The fender imposed too great a duction in performance, and most of these aircraft had it removed and they reverted to the rmal bomber role.

The Ju 88A-15 was a development aimed at increasing the internal bombload of the type to 3000 kg (6,614 lb).

arriage

in wing tanks either side of the engine. The outboard tanks each held 414 litres
board tanks each held 425 litres (93.5 Imp gal). The two internal bomb bays of the
C 50 50-kg (110-lb) bombs. Auxiliary fuel tanks could be fitted in the front (1218 litres/
0 litres/149.6 Imp gal) bomb bays. From the Ju 88A-4 the forward bay was deleted
nent fuel tank. The remaining bay could carry 10 SC 50s, with larger weapons (SC 250,
der the four external wing racks.

Anti-ship weapons

Standard free-fall bombs were the main anti-ship weapon, the
Ju 88 being accurate enough to deliver them with a fair chance
of success. This aircraft carries an SC 500 under the port wing.
Alternatively, mines could be dropped, and this Ju 88 is seen
releasing an LM-B (Luftminen-B) mine. This weapon was
parachute-retarded and the tailcone can be seen detaching from
the main body to deploy the chute. As well as its maritime use,
the LM-B could be used against land targets, exploding on impact
as would a regular bomb. Ju 88s were regularly used to sow
mines around Allied harbours, and also during the D-day landings.

Torpedo bombers

In 1942 a number of Ju 88A-4s were modified to carry
torpedoes, the result being the Ju 88A-4/Torp. A few factory-
prepared torpedo bombers were also produced under the
Ju 88A-17 designation. The two ETC external bomb racks under
each wing were replaced by a single PVC torpedo-carrier, which
could mount an LT F5b torpedo. Each of the weapons weighed
765 kg (1.687 lb). On the starboard side of the nose was a long
bulged fairing that housed the equipment to adjust the steering
mechanism of the torpedoes inflight. The crew was usually
reduced to three and many of the aircraft had the ventral gondola
removed. One Ju 88A-4 was involved in trials of the L 10 glider
torpedo, one of which could be carried on a pylon under each
outer wing panel. Trials were unsuccessful and the scheme
abandoned.

Ju 88A-5

The A-5 was an interim version that introduced the structural improvements of the
Ju 88A-4 while Junkers waited for the supply of the A-4's intended powerplant –
the Jumo 211J – to reach the assembly line. Thus, the A-5 was powered by Jumo
211B-1 or G-1 engines of the Ju 88A-1, yet had the long-span wings and beefed-up
undercarriage of the A-4. It also introduced outer wing hardpoints for the carriage of
250-kg (551-lb) class weapons, but they were rarely used.

Ju 88A-16: conversion trainer version of the A-16, with armament removed.
Ju 88A-17: A-4 modified for the torpedo attack role. It had provision to carry one F5b torpedo
under each wing close to the fuselage.
Ju 88B: revised version of the A-1, with the more streamlined nose that would be a feature of
the Ju 188. Powered by the BMW 801MA radial. V1 followed by 10 **Ju 88B-0s** with lengthened
forward fuselage and long wings of Ju 88A-4. Did not go into production.

*The Ju 88B V1 (above) first flew in early 1940 with an all-new nose on a Ju 88A-1
airframe. Some of the 10 pre-production Ju 88B-0s that followed (below) were used
in the long-range reconnaissance role by the Luftwaffe's HQ reconaissance group.*

Ju 88C-1: heavy fighter version of the aircraft with BMW 801MA radials, based on the V7
prototype. Did not go into production.
Ju 88C-2: conversion of the A-1, with a solid nose and a forward-firing armament of one MG FF
20-mm cannon and three MG 17 7.9-mm machine-guns. Small number produced.
Ju 88C-3: experimental version of C-2 with BMW 801MA engines.
Ju 88C-4: the first heavy fighter variant constructed as such, based on the A-4. Forward-firing
armament increased to three MG FF 20-mm cannon and three 7.9-mm machine-guns.
Ju 88C-5: version of C-4 with BMW 801D-2 engines. Only 10 were built.
Ju 88C-6: first heavy fighter variant produced in quantity. Similar to the C-4, but with additional
armour protection for the crew. Built in large numbers, some served in the night fighter role.

The Ju 88C-6 heavy fighter was employed by both day and night.

Ju 88C-7a: ground attack variant intended for daylight operations, this type carried a forward-
firing armament of two MG FF 20-mm cannon and could carry a bomb load of 500 kg (1,100 lb)
internally in the fuselage.
Ju 88D-0: reconnaissance conversion with Jumo 211B-1 engines.
Ju 88D-1: reconnaissance variant based on the A-4 bomber with Jumo 211J-1/2 engines. The
forward bomb bay was sealed off to house a fuel tank; the rear bomb bay accommodated two or
three aerial cameras. Tropicalised version was designated **Ju 88D-3**.

This Ju 88D served with 2./Aufklärungsgruppe 22 on the Russian front.

Ju 88D-2: interim production version with Jumo 211B-1, G-1 or H-1 engines pending delivery of
Jumo 211J. Tropicalised version initially known as **Ju 88D-2/Trop** but later redesignated as
Ju 88D-4.
Ju 88D-5: version of Ju 88D-2 with standardised three-camera installation.
Ju 88E-0: a Ju 88B-0 was re-engined with BMW 801C engines and fitted with a power turret
above the cockpit housing a single MG 131 machine-gun. Used for Ju 188 development.

This is the power turret and MG 131 13-mm gun as fitted to the Ju 88E-0, which was converted from a B-0. The variant paved the way for the development of the Ju 188.

Ju 88G-1: introduced in the spring of 1944, this night fighter variant was powered by two BMW 801D engines and fitted with an enlarged and squared-off fin and tailplane. The gondola under the nose was removed, and the armament of four 30-mm MG 151 cannon was housed in a tray under the fuselage. This variant was fitted with SN-2 radar and many carried the Naxos and Flensburg systems for homing on emissions from RAF bombers.

This captured Ju 88G-1 carries SN-2 radar, plus FuG 227 Flensburg homing gear.

Ju 88G-4: similar to the G-1, but was fitted with additional equipment including the *schräge Musik* upward-firing cannon. There was also an attachment to the SN-2 radar to enable it to give rearward as well as forwards cover, to give warning of approaching RAF night fighters.
Ju 88G-6: initially similar to G-4 but with BMW 801G engines in the **G-6a** and **G-6b** versions, which differed in equipment. The **G-6c** was powered by Jumo 213As and had the *schräge Musik* installation moved forward.
Ju 88G-7: final night fighter production variant of the Ju 88, this aircraft saw service with the three new types of airborne interception radar used by the Luftwaffe in the final year of the war: the SN-2, the later Neptun and the centimetric wavelength Berlin equipment. Powered by the Jumo 213E.

The Ju 88G-7a had Jumo 213 inline engines, with flame-dampers over the exhausts.

Ju 88G-10: long-range fighter version with lengthened rear fuselage housing additional fuel and Jumo 213A-12 engines. Only a handful built and allocated to Mistel programme.
Ju 88H-1: ultra long-range maritime reconnaissance version of the Ju 88, produced in small numbers to locate convoys far out in the Atlantic. The fuselage was lengthened by 2.29 m (7 ft 6 in) to provide room for four additional fuel tanks. This variant also carried the Hohentwiel ship search radar.
Ju 88H-2: ultra long-range fighter variant with the lengthened fuselage and extra fuel capacity of the H-1, intended to operate against Allied maritime patrol aircraft hunting U-boats in the western Atlantic. In place of the Hohentwiel radar the H-2 carried a nose-mounted battery of six 20-mm MG 151 cannon.
Ju 88H-3/4: further stretch of fuselage with an additional section forward of the wing allowing more fuel to be carried. The angular tail from the Ju 88G was fitted. The H-3 was a reconnaissance aircraft and the H-4 a long-range fighter. Only a few were built and all were diverted to the Mistel programme.
Ju 88P-1: Ju 88A-4 modified to carry a 75-mm PaK 40 cannon, an adaptation of the army's anti-tank weapon, for low-altitude attacks on Soviet tanks. The aircraft also carried additional armour protection for the crew. A few of these aircraft were tested in action, but they were found to be too unwieldy and too vulnerable for general combat operations.

Ju 88P-2: further heavy gun fighter version, fitted with two 37-mm cannon. This variant underwent combat testing in the ground attack role, and also against US heavy bomber formations. The P-2 lacked the performance and the manoeuvrability necessary for these roles, however, and it saw little combat use.
Ju 88P-3: similar to the P-2, but with additional armour protection for the crew.

The Ju 88P-3 carried a pair of hard-hitting 37-mm cannon in the central fairing.

Ju 88P-4: similar to the P-1, but fitted with a single 50-mm cannon.
Ju 88R-1: heavy fighter and night fighter variant, similar to the C-6 but powered by two BMW 801MA radial engines. Built in small numbers.

Now in the RAF Museum, this Ju 88R-1 was captured intact with its FuG 202 radar.

Ju 88R-2: as R-1 except it had BMW 801D engines.
Ju 88S-0: final bomber version of the Ju 88, produced in small numbers. The airframe was cleaned up and it had a more streamlined nose section. Power was provided by BMW 801Ds.
Ju 88S-1: Produxction version with two BMW 801G-2 engines fitted with the GM 1 nitrous oxide injection, which gave it a top speed of 610 km/h (379 mph) at 7924 m (26,000 ft). Defensive armament reduced to one MG 131 13-mm heavy machine-gun firing rearwards from the dorsal position. The crew was reduced to three – pilot, observer and wireless operator.

The Ju 88S-1 dispensed with the gondola, but had a small fairing for the bombsight.

Ju 88S-2: development of S-1 with turbo-supercharged BMW 801TJ engines.
Ju 88S-3: powered by Jumo 213As with GM 1 boosting.
Ju 88T-1: high-speed reconnaissance version of the S-1, with an extra fuel tank in the forward bomb bay and cameras in the rear bomb bay.
Ju 88T-3: Jumo 213-powered equivalent of Ju 88T-1. Production shelved.

This Ju 88T-1 served with 2./Fernaufklärungsgruppe 123 at Athens-Tatoi in 1944.

Right: In common with aircraft in other theatres, Ju 88s in the East often had their national insignia overpainted for night operations. This is a Ju 88A-4 of 7./KG 3.

Below: After the perils of the icy winter, the spring thaw brought a new set of problems for the Luftwaffe in Russia. This forlorn Ju 88 awaits better conditions with an Fw 156, Ju 52 and Ju 86.

Nowhere was the pace of operations more severe than during the operations to capture the Soviet naval base at Sevastopol in the summer of 1942. The bomber ace Major Werner Baumbach, commander of KG 30, flew on some of the attacking aircraft to observe conditions in the area. He later wrote in his autobiography (published in English as *Broken Swastika*, Robert Hale).

"From the air Sevastopol looked like a painter's battle panorama. In the early morning the sky swarmed with aircraft hurrying to unload their bombs on the town. Thousands of bombs – more than 2,400 tons of high explosive and 23,000 incendiaries – were dropped on the town and fortress. A single sortie took no more than 20 minutes. By the time you had gained the necessary altitude you were in the target area . . . the Russian AA was silenced in the first few days so the danger to aircraft was less than in attacks on the Caucasus harbours or Russian airfields. Yet our work at Sevastopol made the highest demands on men and material. Twelve, 14 and even up to 18 sorties were made daily by individual crews. A Ju 88 with fuel tanks full made three or four sorties without the crew stretching their legs."

It should be noted that the Luftwaffe had no system of operational tours with a fixed number of combat missions, as in the RAF or the USAAF. Luftwaffe crews remained in action continually, their only breaks from combat flying being during their limited periods of leave, or when their unit was withdrawn to re-form or re-equip with a new aircraft type or subtype. Thus many aircrews became worn out, both mentally and physically.

Fighter operations

The Ju 88Cs of I./NJG 2 flew almost nightly intruder patrols over airfields in England, until October 1941. Then the demands of the Eastern Front meant that the Luftwaffe was seriously over-extended, and the unit moved to conventional long-range day fighter operations. Ju 88C production continued at a low rate, and during the whole of 1941 only 65 were built.

The first heavy fighter version of the Ju 88 to be built as such 'from the ground up' was the C-4, which entered production in 1941. The new variant had a new wing with a span increased from 18.26 m (59 ft 11 in) to 20.08 m (65 ft 10½ in), there was increased armour protection for the crew and it had no provision to carry bombs internally. It was followed in

The solid-nosed Ju 88C played a significant role in the Russian front fighting, being used for long-range fighter, low-level strafing and bombing attacks. They targeted the Soviet road and rail networks in particular. Above is an aircraft of KG 3, while the hastily camouflaged aircraft at right was flown by KG 76 in May 1943. Like the latter example, many Ju 88Cs flew without the fuselage gondola installed.

Wearing a white distemper winter camouflage, this Ju 88C-6 of 4./KG 76 has had fake 'beetle's eye' glazing painted on in an attempt to fool Soviet fighter pilots into thinking the aircraft is a bomber version.

production later in the year by the C-6, with a forward-firing armament of three MG FF 20-mm cannon and three 7.9-mm machine-guns.

During 1942 the first German airborne interception radar, the Lichtenstein BC, entered service in quantity. The equipment had a maximum range of 4 km (2½ miles) and a minimum range of 600 ft (183 m). At first the radar was not popular with night fighter crews. The equipment had its share of teething troubles, and the drag from the nose-mounted aerial array clipped about 8 km/h (5 mph) off the fighter's top speed. Only when the RAF extended its operations into moonless nights, later in 1942, was the value of the radar appreciated. At the same time the set's reliability improved and it became a standard fit on all Luftwaffe night fighters.

Until the summer of 1943 the Ju 88 formed a small part of the home defence night fighter force, equipping only one gruppe and partially equipping another. During that autumn the force abandoned its close control tactics, and sought to concentrate fighters to establish long-running battles with the raiding bomber streams. The Messerschmitt Bf 110, previously the mainstay of the night fighter arm, gave way to the Ju 88 that was much better suited to the task. Not only was it faster, but it had a longer range and was better able to carry the greater weights of armament and electronic systems then required. In the course of 1944 the Ju 88 replaced the Bf 110 in most night fighter gruppen.

A night fighter pilot remembers

On completion of his training as a night fighter pilot near the end of 1943, Unteroffizier Emil Nonenmacher went to III. Gruppe of Nachtjagdgeschwader 2 based at Schiphol in Holland flying Messerschmitt Bf 110s. Soon afterwards the unit converted on to the Ju 88C. Initially the change was greeted with alarm, as Nonenmacher recalled:

"I and a lot of other pilots were apprehensive about converting to the Ju 88 – it was rumoured to be a dangerous aircraft in the hands of an inexperienced pilot. In the event our fears proved groundless, the conversion was painless. I was very pleasantly surprised with the Ju 88's handling characteris-

Below: Many night fighters, like this Ju 88C-6, had two MG 151/20 cannon firing obliquely upwards in the schräge Musik *installation.*

Junkers Ju 88s in service with operational units, 17 May 1943

This section gives the strength of Junkers Ju 88 front-line units on 17 May 1943. On that day the front-line units reported a total of 1,125 Ju 88s (of which 718 were serviceable) on strength. The majority, 761 aircraft, served with bomber units. Yet of the 26 bomber gruppen equipped with the type, seven – or nearly one-quarter of the gruppen – had been temporarily withdrawn from operations to reform after suffering heavy losses. Also on that day, the long-range reconnaissance units reported 223 Ju 88s on strength.

The first figure gives the total number of aircraft, the second figure gives the number serviceable.

BOMBER UNITS

Lehrgeschwader 1		
Stab	1	1
I. Gruppe	37	29
II. Gruppe	31	10[1]

Kampfgeschwader 1		
Stab	4	4
I. Gruppe	20	0[1]
II. Gruppe	26	14
III. Gruppe	37	18

Kampfgeschwader 3		
Stab	1	0
I. Gruppe	37	29[1]
II. Gruppe	37	27
III. Gruppe	31	13

Kampfgeschwader 6		
I. Gruppe	31	21[2]
II. Gruppe	20	15[2]
III. Gruppe	34	28[2]

Kampfgeschwader 30		
I. Gruppe	37	32
II. Gruppe	26	14
III. Gruppe	32	30

Kampfgeschwader 51		
I. Gruppe	37	28
II. Gruppe	26	14
III. Gruppe	21	11

Kampfgeschwader 54		
Stab	1	1
I. Gruppe	20	11
II. Gruppe	22	10
III. Gruppe	34	16[1]

Kampfgeschwader 76		
Stab	2	2
I. Gruppe	36	21
II. Gruppe	5	31
III. Gruppe	32	23

Kampfgeschwader 77		
II. Gruppe	26	20[1]
III. Gruppe	20	14

Luftwaffe Stab Kroatian		
Einsatzstaffel	37	32

ANTI-SHIPPING UNITS

Kampfgeschwader 26		
III. Gruppe	13	7[3]
Beleuchterstaffel	5	0[4]

LONG RANGE FIGHTER UNITS

Zerstörergeschwader 26		
10. Staffel	12	12

Kampfgeschwader 40		
V. Gruppe	37	28

NIGHT FIGHTER UNITS

Naohtjagdgeschwader 2		
Stab	1	0
I. Gruppe	18	8
II. Gruppe	11	10

Nachtjagdgeschwader 3		
IV. Gruppe	25	22

Nachtjagdgeschwader 5		
IV. Gruppe	15	11[5]

Zerstörergeschwader 1		
Nachtjagd Schwarm	4	2

RECONNAISSANCE UNITS

Aufklärungsgruppe 11		
4. Staffel	6	4

Aufklärungsgruppe 14		
4. Staffel	12	10

Aufklärungsgruppe 22		
1. Staffel	7	5
2. Staffel	12	8
3. Staffel	12	9

Aufklärungsgruppe 33		
1. Staffel	12	7
3. Staffel	12	8

Aufklärungsgruppe 100		
1. Staffel	10	6[5]
2. Staffel	12	8

Aufklärungsgruppe 120		
1. Staffel	11	5

Aufklärungsgruppe 121		
3. Staffel	12	6
4. Staffel	9	6

Aufklärungsgruppe 122		
1. Staffel	9	2[5]
2. Staffel	6	2[5]
3. Staffel	11	5
4. Staffel	12	9
5. Staffel	10	7

Aufklärungsgruppe 123		
1. Staffel	12	6
2. Staffel	12	7
3. Staffel	12	6

Aufklärungsgruppe 124		
1. Staffel	12	11

Versuchsverband der Oberkommando der Luftwaffe		
1. Staffel	4	3[6]

Sturzkampfgeschwader 2		
Stab	6	6[7]

1. Unit reforming.
2. Unit forming
3. Torpedo-bomber unit
4. Unit to illuminate enemy shipping, to facilitate night torpedo attacks
5. Unit operated other type or types
6. Reconnaissance unit
7. Unit operated Ju 88s for pre- and post-strike reconnaissance

Above and right: The Ju 88Cs of V./KG 40 flew long-range fighter missions, mainly over the Bay of Biscay. The Bay was a favourite hunting ground of Allied maritime patrol aircraft in the war against the U-boats, and the Ju 88s were employed against them. The unit was later redesignated as I. Gruppe of Zerstörergeschwader 1, but its role remained the same.

Several aircraft were modified to the Ju 88A-6/U configuration, an anti-shipping specialist with FuG 200 Hohentwiel radar, a three-man crew and the ventral gondola removed. This example operated over the Mediterranean from southern Italy in 1943.

tics. At first it was a bit disconcerting to be sitting on the left side of the aircraft, instead of in the centre as in the Me 110, but one soon got used to that. The view to the right was not good, but that was not really necessary for night fighting. The Ju 88 was a bit heavier on the controls than the 110, but it was more stable and far better for instrument flying. If properly trimmed, it would fly hands off. A good pilot could do aerobatics in the 88 as well as in the 110. And the 88 was a bit faster, 25 km/h [15 mph] low down and 40 km/h [25 mph] high up."

Radio spoofing

Emil Nonenmacher flew a night interception mission on the night of 14/15 January 1944, when a force of 496 RAF bombers set out to attack Brunswick. For III./NJG 2 the night's operations were not a success:

"The information on the whereabouts of the enemy came in too late. Also our running commentary was badly jammed by the enemy, with a continual 'Dodle Dodle Dodle' on our radio channels. Then came a call 'All Dromedaries [the radio callsign of NJG 3] and Cockatoos [NJG 2] return immediately to base, on account of ground fog.' That call was unjammed – because it came from the enemy! I did not believe it, the caller did not use the proper code word, but several other pilots obeyed it and went home. There was a big investigation afterwards."

Above: Another FuG 200-equipped Ju 88A-6/U shows the crew access hatch and ladder. Ju 88s were heavily involved in the anti-shipping role in the Mediterranean and Norway, and to a lesser extent from France, with Kampfgruppe 606 operating the type from Brest.

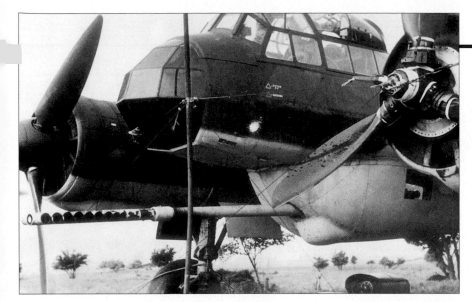

Ju 88P – 'big-gun' 88s

The proliferation of Red Army tanks in 1942 was the spur to development of the Ju 88P series, which carried large-calibre anti-tank guns under the belly, although they were later tested against both day and night bombers. Small numbers saw action on the Russian front with Panzerjäger-staffel 92. The Ju 88P generally proved slow, unwieldy and easy to intercept in the air, and the low rate of fire of the big guns reduced its combat effectiveness.

Above: Known as 'die dicke Bertha' (Fat Bertha), the Ju 88P V1 was a Ju 88A-4 converted to carry a 75-mm KwK 39 hand-loaded anti-tank cannon. The barrel was slightly depressed to allow attacks from a shallow dive. 'Production' Ju 88P-1 conversions had a solid nose and the better PaK 40 cannon of similar calibre.

Right: A number of 'big-gun' aircraft were tested in the night fighter role, including this Ju 88P-3. It had a pair of moveable MK 103 30-mm cannon in the fairing, but they slowed the aircraft too much. The three pipes under the fuselage created a loud and distinctive note to enable ground posts to track the aircraft at night.

Other units did not react to the spoof, and had greater success. That night the RAF lost 38 bombers, 7.6 per cent of the force.

Emil Nonenmacher gained his first aerial victory one week later, on the night of 21/22 January when 648 RAF bombers attacked Magdeburg. After taking off from Gilze-Rijen he was told the target was Berlin. He headed in that direction in the climb and on the way he listened to the German controller's broadcasts, and to the British attempts to jam them: "It was a bit wearing, but one got used to it."

As the Ju 88 passed over Magdeburg he heard that the city had been attacked, and he could see the glow of the fires on the ground. It seemed they had got there too late to engage the raiders. Nonenmacher turned for home.

"About 10 minutes after we left Magdeburg, I caught sight of an aircraft about 400 metres ahead and slightly above. In the darkness it looked just like an Me 110, from the indistinct silhouette we could make out the twin fins and rudders. It seemed we would have company to fly home with, so I moved into formation just off its right wing. Then, as we closed in, I

The Ju 88P-2 and P-3 (with better armour) had a pair of BK 3,7 (Flak 38) 37-mm cannon offset to port. The size of the ventral fairing is all too apparent in this shot of a Ju 88P-3 (with a Ju 290 in the background), and its affect on performance can easily be imagined.

Converted from a Ju 88R-2, the V58 (right) was the prototype for the Ju 88G series. It was armed with six 20-mm MG 151s, four of which were carried in the Waffentropfen ventral tray (above). The two starboard nose cannon were deleted from production Ju 88G-1s.

Death of a night fighter ace

In the third week of January 1944, the top-scoring Luftwaffe night fighter ace was Major Prince Heinrich zu Sayn Wittgenstein, holder of the Knight's Cross with Oak Leaves and Kommodore of Nachtjagdgeschwader 2. When he took off in his Junkers Ju 88 on an interception sortie after dark on 21 January, his score stood at 79 night victories.

That night 648 bombers set out to bomb Magdeburg and, as the raiders swept in over northern Germany, Wittgenstein located the bomber stream and was soon in the thick of the fighting. Afterwards his radar operator, Feldwebel Ostheimer, described what happened:

"At about 10 pm I picked up the first contact on my [SN-2] airborne radar. I gave the pilot directions and a little later our target was seen: it was a Lancaster. We moved into position and opened fire, and the aircraft immediately caught fire in the left wing. It went down at a steep angle and started to spin. Between 10 and 10.05 pm the bomber crashed and went off with a violent explosion. I watched the crash.

"Again we searched. At times I could see as many as six aircraft on my radar. After some further directions, the next target was in sight – another Lancaster. After the first burst from us there was a small fire, and the machine dropped back its left wing and went down in a vertical dive. Shortly afterwards I saw it crash. It was some time between 10.10 and 10.15 pm. When it crashed there were heavy detonations, most probably it was the bomb load. After a short interval we again saw a Lancaster. After a long burst of fire the bomber caught fire and went down. I saw it crash some time between 10.25 pm and 10.30 pm. The exact time is not known. Immediately afterwards we saw yet another four-engined bomber. We were in the middle of the 'bomber stream'. After one pass this bomber went down in flames; at about 10.40 p.m. I saw the crash."

Within a few minutes, Ostheimer had another target blip on his radar. After a few course alterations, yet another Lancaster hove into sight. Wittgenstein closed in and delivered an attack, and the bomber started to burn. The fire went out, so the German pilot moved into position for another attack. Suddenly a series of explosions rocked the Junkers Ju 88, and its left wing caught fire. Ostheimer's account continued:

"As I saw this, the canopy above my head flew off and I heard on the intercom a shout of 'Raus!' [Get out!] I tore off my oxygen mask and helmet and was then thrown out of the aircraft."

Ostheimer landed safely by parachute, but Wittgenstein was unable to escape from his cockpit. His body was later found in the wreckage. Only the

Luftwaffe night fighters were also used occasionally in the desperate attempts to halt USAAF day raids. Here a Ju 88G comes under attack. Such encounters brought back the first evidence that the Luftwaffe had introduced a new type of radar – the Lichtenstein SN-2.

night before, the intrepid pair had gone so close below a Halifax to make certain of hitting it, that the plunging bomber had struck their aircraft. They had been lucky to escape with their lives. This time the night fighter ace had not been so fortunate.

Almost certainly the Ju 88 had been attacked by another bomber in the stream, though no RAF claim links with the loss. It seems that Wittgenstein had fallen victim to a surprise attack from below – a form of attack that he himself had used to great effect. On the night of 21/22 January, 55 RAF bombers failed to return.

Right: A Ju 88G-1 has its compass swung. Early Ju 88Gs were powered by the BMW 801 radial.

Below: On 13 July 1944 this Ju 88G-1 from 7./NJG 2 landed by mistake at RAF Woodbridge, bringing with it the secrets of the Lichtenstein SN-2 radar and the FuG 227 Flensburg device that homed on Monica emissions.

began to get an uneasy feeling that something was not right. Suddenly I caught a glimpse of some Plexiglas to the rear of the rudders, and some more mid-way along the fuselage. I must have been flying happily alongside the aircraft for about 20 seconds before I finally realised: it was a four-engined bomber. A Halifax! An enemy!

"But if we had been taken aback, those in the bomber must have been asleep because they continued straight ahead. I throttled back and pulled away from him, then slowly moved into a firing position behind him. I closed to within 50 metres [164 ft], about half the ring of my reflector sight was filled with the bomber's rear turret. I can still see the picture of the sight exactly on that turret with the guns pointing upwards.

Ju 88 night fighter equipment

FuG 202/212 Lichtenstein BC/C-1

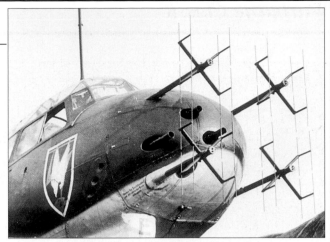

Germany's first airborne intercept radar, the FuG 202 Lichtenstein BC (code-named 'Emil-Emil'), was developed by Telefunken and entered service in 1942. It used the *Matratzen* (mattress) array of four dipole arrays. The standard Ju 88 installation is seen on a Ju 88R-1 (right, with *Englandblitz* badge) and in the unusual installation in the Ju 88 V58 (left). FuG 212 Lichtenstein C-1 was redesigned internally but used the same array. The radar displayed its returns on three cathode ray tubes in the cockpit.

FuG 220/228 Lichtenstein SN-2/SN-3

Another Telefunken development, the FuG 220 employed four *Hirschgeweih* (antlers) aerials and was widely carried by Ju 88Gs in several forms. The first Lichtenstein SN-2b installations also employed an FuG 212 to cover the ranges below which the basic SN-2 could not operate. These used a full *Matratzen* array (as seen at left), or more usually a single *Weitwinkel* (wide-angle) pole. Improvements to the FuG 220 allowed the deletion of the FuG 212 in the SN-2c installation (above left), which was followed by the SN-2d with dipoles angled at 45° (above right). At least one Ju 88 had its SN-2 dipoles mounted above and below the fuselage. The FuG 223 Lichtenstein SN-3 used blade aerials, although only 10 are believed to have been delivered.

FuG 218 Neptun VR/GR

A product of Flugfunk Forschungsanstalt (FFO) and Siemens, the FuG 218 Neptun VR was a combined airborne intercept/tail-warning set, with an antenna in the nose and another mounted at the top of the tail. The nose antenna consisted of either a four-dipole set on a single central pole, or a *Morgenstern* (morning star) arrangement that was either covered (right) or uncovered (left, on a Ju 88G-6 from Stab II./NJG 5). Neptun GR had boosted power and six selectable frequencies. The Morgenstern antenna has also been described as being an experimental array for the FuG 220.

FuG 240/1 Berlin N-1a

Developed by Telefunken from a captured RAF H₂S set, the centimetric Berlin radar was fitted to only 10 Ju 88s before the war's end, including this Ju 88G-7c of II./NJG 4. Operating in the 9- to 9.3-cm wavelength band and with a frequency of around 3 GHz, the Berlin was virtually immune to the effects of Window (chaff). The dish antenna was mounted behind a simple wooden nosecone.

FuG 350 Naxos Z/ZR

FuG 350 was developed by Telefunken as a means of homing in on the emissions of the H₂S and other Allied centimetric radars, although it only gave bearing information. It was produced in many forms, the most common being the basic Naxos Z mounted in a fairing above the cockpit (left, on an NJG 102 Ju 88G). Some Ju 88s had the Naxos ZR installation (right), which combined Naxos Z with a tail-warning capability.

A Ju 88A-4 of KG 76 flies over the Alps in early 1944. The aircraft was deploying to France to join the bomber force being assembled for Operation Steinbock, the Luftwaffe's last major offensive bombing campaign of the war.

themselves to the sight in front of me. Then I could make out the flashes of my cannon shells exploding against the bomber, and others ricocheting in all directions off the structure. Then I finally remembered to take my thumb off the firing button."

Nonenmacher opened the range to see what would happen next. He expected to see the bomber to burst into flames, or blow up, or do something. But it was still in front of him, flying straight and level. It was as though the burst of cannon fire had achieved nothing at all.

"So I moved in once more and fired a second burst. And still the bomber flew on without catching fire. This time I was prepared for the bright muzzle flash. By screwing up my eyes a little I could make out my shells hitting the bomber or, in many cases, ricocheting off it. So I moved in for a third burst. But my first two bursts had been too long, and after a few rounds all my guns jammed. Again I pulled away a little and I could see the bomber was beginning to go down, though only slowly and there was no fire. Gradually the bomber's descent became steeper and steeper and it entered cloud going almost vertically. Then, through the cloud layer, we saw the glow as the bomber exploded when it hit the ground."

Emil Nonenmacher ended the war credited with six night victories, all achieved while flying the Ju 88.

There was no need to make any allowance for deflection. I pressed the firing button – and nothing happened!"

In the excitement of discovering that this was an enemy bomber, the flight engineer had forgotten to switch on the compressed air to cock the guns. Nonenmacher shouted to him to switch on the compressed air while he held his sight on the bomber.

"Still the rear gunner had not noticed me. I pressed the firing button a second time, and this time the armament functioned properly. During my time at gunnery school I had got good scores, but I had never fired at night before. I was so busy concentrating on the sight that the unexpectedly bright muzzle flash blinded me almost completely. I was so surprised that I forgot to release the firing button. As a result I fired for about 5 seconds, about twice as long as I should have done. Only after about 4 seconds did my eyes begin to accustom

Demise of the Luftwaffe bomber arm

At the beginning of 1944 the types planned to replace the Ju 88, the Junkers Ju 188 and the Heinkel He 177, were slow in coming into service. As a result, the Ju 88 still made up the backbone of the Luftwaffe bomber arm. In the third week in January, the Luftwaffe assembled a large force for Operation Steinbock, a renewal of the attacks on London and other cities in England. Of the 524 bombers earmarked for that operation 290, or more than half, were Ju 88s.

A few of these Ju 88s were of the new Ju 88S variant, with a much cleaned up airframe, serving with the pathfinder unit I./KG 66. Power was from two BMW 801G engines with nitrous oxide injection, which boosted the maximum speed to 610 km/h (379 mph) at high altitude. Releasing their markers from 30,000 ft (9144 m) or so over the target, these aircraft made difficult targets for the defenders. They suffered few losses.

'Black men' – as ground crew were known in Luftwaffe parlance – load a massive AB 1000 container on to a KG 76 Ju 88 for a Steinbock raid on Britain. The AB 1000 contained 620 1-kg incendiary munitions. After the container was dropped, the incendiaries were released at pre-set intervals to produce a series of dense bomb clusters scattered over a wide area.

The great majority of the Ju 88s taking part in Steinbock were the elderly A-4 version, however. The performance of this variant had improved little since it entered service in 1940, and by 1944 it was obsolescent. These outdated bombers suffered heavily at the hands of the high-performance Mosquito night fighters confronting them.

Operation Steinbock petered out in May 1944. Thereafter the Luftwaffe bomber force had only a couple of weeks to lick its wounds before its next, and final, great exertion: the operations against the long-expected Allied invasion of France. The large anti-shipping force assembled in the south of that country to counter the landings included two gruppen of KG 30 and two from KG 77, with a total of 125 Ju 88A-17 torpedo-bombers. This variant, a development of the obsolescent A-4, carried one F5b torpedo on each inboard wing section close to the fuselage.

As night fell on D-day, 6 June, the Ju 88 torpedo-bombers set out *en masse* to strike at the huge concentration of shipping supporting the invasion. In addition there were 130

Operation Steinbock – the 'little Blitz'

Above and right: One of the key units in Steinbock was I./KG 66, which flew the Ju 88S-1 in the pathfinder role. The S was a belated attempt to increase the performance of the standard Ju 88 bomber. The most obvious feature was the smooth glazed nose. Power for the S-1 came from a BMW 801G-2 radial with GM 1 nitrous oxide boosting. The S-2 had a turbo-supercharged BMW 801TJ without GM 1.

sorties by conventional bombers, a large proportion of them Ju 88s. That night there were targets aplenty off the coast of Normandy: seven battleships, 23 cruisers, 105 destroyers, more than a thousand smaller vessels and more than four thousand landing craft supported the landings. Yet the attacking aircraft faced insurmountable problems. During their approach to and return flights from the beachhead the attackers suffered casualties not only from Allied night fighters, but also from German flak units long accustomed to regarding any aircraft that came within range as hostile. Those bombers that reached the invasion area encountered such a violent reception from guns ashore and afloat that many attacks had to be broken off prematurely. Not a single ship was lost to air attack on that first night.

Mining the shallows

During the next 10 nights the Luftwaffe units involved in the anti-shipping operations suffered further losses, and were able to sink only two destroyers and four smaller ships. Finally the Ju 88 units gave up trying to make direct attacks on the ships. Instead they dropped large numbers of pressure mines in the shallow waters off the Normandy coast. Detonated by the pressure change from the wake of a passing ship, the new type of mine caused much inconvenience. They were difficult to sweep, and could be countered only by ordering ships to reduce speed to a minimum while moving through shallow water. Yet while the mines caused difficulties, delays and a few losses, they could not be decisive.

Once the Allied armies were safely ashore in Normandy, the Allied heavy bomber fleets turned their attention to the destruction of the German oil industry. The effect was immediate and devastating. Compared with 175,000 tons of aviation fuel produced in April 1944, the industry turned out only 55,000 tons in June. By early July the fuel crisis was starting to bite and the Luftwaffe had to curtail or cease operations by its multi-engined bombers. One by one these units withdrew to Germany, where many units disbanded. The Ju 88-equipped reconnaissance units soldiered on somewhat longer, though they were limited to much reduced sortie rates.

The final chapter in the Ju 88 story began in the summer of 1944, when a few hundred of these airframes were converted into flying bombs under the Mistel (mistletoe) programme. The aircraft had its nose and the crew compartment removed.

Beethoven project – the Mistels

Beethoven was the codename chosen for a programme that used Ju 88 airframes to act as flying bombs, flown to and aimed at their targets by a piloted fighter mounted piggyback. Once the bomber was lined up, the pilot engaged the bomber's autopilot, and then fired explosive bolts to allow him to fly free.

Above: The Mistel 2 combination paired a Ju 88G-1 with an Fw 190A or F. This II./KG 200 Mistel 2 was captured intact at Merseburg, where it had deployed for Operation Eisenhammer, a major attack against the Soviet armament industry which, in the event, was never launched.

Left: The Mistel 1 was a Bf 109F mounted on top of a Ju 88A-4. A simple nose was fitted for ferrying.

Below: This is a fully operational Mistel 1, with the warhead installed. It was the only combination to be used operationally.

In its place a huge 7,800-lb (3538-kg) shaped-charge warhead was fitted on the nose. A Bf 109 or an Fw 190 fighter was rigidly mounted on top of the bomber, and both aircraft were controlled from the cockpit of the fighter. The pilot of the latter flew the Mistel to the target area and aligned it on the target, then fired explosive bolts to separate the fighter from the Ju 88. The Ju 88, now flying on autopilot, flew straight ahead and exploded when it impacted the target. Meanwhile, the fighter pilot headed away from the target and made for home at top speed.

The Mistel enabled the Luftwaffe to deliver a highly destructive explosive charge against a high-value target, a capability it had not possessed before. Some imaginative operations were planned for the Mistel, including a large-scale attack on the British Home Fleet at Scapa Flow, and another against the electrical power generation system in the Soviet Union. Neither

operation took place, and many of the Misteln prepared for these attacks were used in attempts to wreck bridges carrying advancing Soviet forces over river barriers during the final stages of the war. The Mistel lacked the accuracy necessary to engage such small targets, however, and it had little success in this role.

On 9 April 1945, the final date for which reasonably detailed figures exist, there were about 350 Ju 88s serving with front-line night fighter units. A few more served with long-range reconnaissance units. By that stage of the war there was no front-line bomber unit operating the Ju 88.

The Ju 88 summed up

In 1939 the Junkers Ju 88 represented the cutting edge of bomber design and it was, arguably, the best all-round medium bomber in the world. For that time it had an excellent performance and it was highly manoeuvrable. It carried a reasonably large bomb load, which it could deliver with great accuracy in a diving attack. Allied test and combat reports written during the war make it clear that the Ju 88 made a profound impression on Germany's enemies.

Yet in wartime such advantages could be sustained only if a new type, or a modified aircraft with a significantly improved performance, became available in quantity and in time to replace it. That did not happen in the case of the Ju 88 in the bomber role. By the middle of the war the type was obsolescent. At the very time when the Allied fighter and AA gun defences were becoming progressively more potent, the elderly Ju 88A-4 and its derivatives lacked the performance to avoid these threats. The inevitable result was that Ju 88 units suffered increasingly heavy attrition on all fronts, and lost much of their ability to destroy targets. The high-performance Ju 88S variant appeared too late, and in too small numbers, to change matters.

During 1943 the Luftwaffe was forced on to the defensive over the Fatherland, and the Ju 88 was pressed into service in large numbers in the night fighter role. By the end of 1944 it equipped the bulk of the night fighter force.

Ju 88 production ended in May 1945, as Allied troops overran the factories building them. By then the Luftwaffe had taken delivery of more than 14,500 Ju 88s of all variants. It was one of the great combat aircraft types of all time, and it served in the front-line from the first day of the war until the last.

Dr Alfred Price

Ju 88s in foreign service

Finland

In April 1943 the Finnish air force acquired 24 Ju 88A-4s (serialled JK-251 to 274). The aircraft were put into service with Pommituslentolaivue 44 within Lentorykmentti 4 at Onttola. They were put into use immediately against Soviet forces, impressing the Finns with their bombing accuracy. They were a vital part of the small force which halted the Soviet advance along the Karelian isthmus in 38 days of hectic fighting in June/July 1944, bringing about an end to the Continuation War and resulting in an armistice signed on 4 September.

Among the truce conditions, which were formalised on 19 September, was the need to expel all German forces from Finnish territory. So began the Lappland War, in which Finnish combat units pushed German forces back into Norway. On 4 December the Finnish air force was partially immobilised and reorganised, the Ju 88 unit becoming PLeLv 43. By January 1945 it was the only combat unit still active in Lappland, flying on missions until the German expulsion campaign was successfully completed on 27 March. Along with other bomber units, PLeLv 43 was disbanded soon after the war ended.

France

In the vacuum created by the German departure from southern France a number of units were formed using available equipment. Among them was Groupement Dor, which formed in the Forces Françaises de l'Intérieur on 4 September 1944. It operated

Ju 88A-4s gathered from various sources and overhauled by SNCASE at Toulouse. It flew its first combat missions on 16 October against German positions in the Gironde estuary. Soon after the unit became part of the Forces Françaises de l'Atlantique and was redesignated as

Groupe de Bombardement I/31 'Aunis'. From March 1945 the output of aircraft from SNCASE was augmented by that from Ateliers Aéronautiques de Boulogne. The following month GB I/31 ceased operations, having received 22 aircraft. In the post-war era the group continued to fly the bomber until 1947,

when the survivors were transferred to second-line duties such as navigation and gunnery training at Cazaux. AAB produced a few more aircraft, and some ex-Luftwaffe Ju 88D and T recce aircraft were taken on charge after refurbishment. French Ju 88s remained in use into the 1950s.

This semi-derelict aircraft is a Ju 88D-5 reconnaissance aircraft, a few of which were used for some years by the post-war Armée de l'Air.

Seen receiving attention is a rudderless Ju 88T, one of a number of ex-Luftwaffe aircraft taken over by the French air force.

Hungary

Hungary's pro-Axis air arm received around 40 Ju 88A-4 bombers (below right), which served with the 102/2 Fast Bomber Squadron, among others.

Hungary also received numbers of the Ju 88D-1 reconnaissance aircraft (serving with I Long-Range Reconnaissance Squadron) and the Ju 88C (below, with fake glazing).

Italy

In early 1943 the Regia Aeronautica began to receive 52 Ju 88s – mostly A-4s but with a few D-1 reconnaissance aircraft. The bombers were allocated to 212ª and 213ª Squadriglia of the 51º Gruppo

Bombardamento Terrestre, part of the 9º Stormo at Viterbo. The recce aircraft went to the 172ª Squadriglia Autonoma within the the 9º Stormo. While the units were still working up Italy capitulated on 8 September 1943, ending the Ju 88's brief career in Italian hands.

Romania

The Royal Romanian Air Force acquired the Ju 88A-4 in 1943 to form the 5th Bomber Group within Corpul 1 Aerian. The group comprised three squadrons (Escadrile 75, 76 and 77) flying the Ju 88A-4. The new group was deployed to Zaporozhye in the Ukraine in the summer, fighting alongside Luftwaffe aircraft on the southern sector of the Russian front. The aircraft pulled back to Romania in the face of Soviet advances.

Following the Romanian coup on 23 August 1944, the country turned on its former ally and the Ju 88s of Grupul 5 were used against German forces in the Klausenburg region.

Romania also received Ju 88Ds which flew with the 2nd Squadron of the Long-Range Reconnaissance Wing.

Romanian Ju 88A-4s are seen operating from a grass field in the Ukraine in the summer of 1943.

Flying from ship and shore, the US Navy was heavily embroiled in the fighting in Korea. The service was also in the process of a significant transformation, so that its combat inventory represented a fascinating mix of first-generation jets and World War II-era props.

US Navy in Korea

Above: Skyraiders of VA-115 strike inland on a mission from USS Philippine Sea (CV 47). Throughout the war US Navy attack aircraft maintained a continuous assault against North Korean forces. Skyraiders operated in strategic attack, interdiction and close support roles.

Left: USS Kearsarge (CV 33) prepares to launch its air wing in 1952. The F2H-2P Banshees are from VC-61, the fighter Banshee from VF-11, and the Corsairs are from VF-884. The bomb-laden Panther is from VF-721.

Below: USS Antietam (CV 36) sails in placid waters with its full air wing aboard. Following Korean service, the carrier entered the yard to be modified with a trial angled deck, the first carrier to have this innovation. As well as Task Force 77's fast-carrier 'Essex' boats like Antietam, mention must also be made of the hard-working escort carriers of TF 86, notably Sicily (CVE 118) that undertook four combat cruises.

Above: Numerically the most important Navy type was the elderly Vought F4U Corsair, each carrier usually embarking two squadrons of 16 aircraft each. They were tasked with close air support, and proved to be effective in this role, even though it was dangerous. By the war's end the US Navy had lost 312 Corsairs (compared with 124 ADs and 64 F9Fs). A small number of F4U-5Ns were deployed ashore in the night-fighter role towards the end of the war to counter North Korean 'Bedcheck Charlie' nuisance raiders. VC-3's Lt Guy Bordelon, operating as part of Detachment Dog at Pyongtaek (K-6) shot down five of them to become the only non-Sabre ace of the war. This photograph captures a VF-871 Corsair on USS Essex (CV 9) in 1952.

Providing some insight into the dangers associated with operations from straight-deck carriers, a Corsair can be seen above landing on Valley Forge (CV 45, above), while below are VF-653 aircraft returning to the same carrier. The 1950 line-up (left) is at Iwakuni, from where the Corsairs will be flown to the carrier and thrown into the hectic fighting on the Korean peninsula.

The Grumman Panther was the most prevalent Navy jet in Korea, serving in F9F-2, -3 and -5 fighter versions. Each carrier usually had one squadron of 18 aircraft embarked. A few F9F-2P and -5P reconnaissance aircraft also saw action. Panthers from VF-51 scored the Navy's first kills of the war (Yak-9s) on 3 July 1950, while VF-111 claimed the first jet kill against a MiG-15 on 9 November. A further five kills were officially credited, with several more probables. The straight-wing Panther could not compete on equal terms with the swept-wing MiG-15, and was consequently used mainly for close support missions. When engaged, though, it could put its heavy-hitting 20-mm cannon to good use. Above are Panthers from VF-721, preparing to launch from USS Boxer (CV 21) in March 1951. At left is a pair of F9Fs from VF-111 at a shore base. The central aircraft has had its rudder removed.

Above: An F9F-2 of VF-191 is parked on the pierced steel planking at Taegu in 1952.

Right: This F9F-2 belonged to VF-721, and is seen on the wooden deck of Boxer in 1951. The vessel completed four Korean war cruises.

Below: The lack of armament suggests that this Panther is one of the few F9F-2P photo-recon conversions.

Above: The McDonnell F2H Banshee was not used in Korea in the same numbers as the Panther, being deployed aboard carriers in detachments. The F2H-2 version was mainly used to provide fighter escort for the 'long-nose' F2H-2P photo-reconnaissance platform. Here both versions are seen on the deck of USS Essex in Yokosuka harbour.

Right: Receiving attention on the dirt at Pohang is a McDonnell F2H-2 of VF-172. Although Navy fighters did occasionally operate from land bases in Korea, they were mainly seen ashore as a result of emergency landings after developing problems or taking hits while in combat 'up north'.

Below: The 'PP' codes identified Composite Squadron VC-61, which provided photo-reconnaissance detachments to carriers operating off Korea. This F2H-2P Banshee is seen at Kangnung (K-18) in 1953.

USS Essex (CV 9) *is seen above in port at Yokosuka in December 1951 with Carrier Air Group Five aboard. The carrier was in the middle of a combat deployment at the time. VF-54 Skyraiders and F4U Corsairs are seen (left) being refuelled aboard the vessel in early 1952. The 'Essex' class bore the brunt of carrier operations off Korea, the lead ship deploying twice. Most of the class that were involved in combat operations had received the SCB-27A modifications that beefed up the deck, lifts and catapults to handle jet aircraft.*

In various forms the Douglas AD Skyraider – the 'Able Dog' or 'Spad' – became the backbone of the US Navy's attack efforts in Korea, alongside the venerable F4U Corsair. Skyraider attack squadrons undertook 24 combat cruises during the conflict. As well as day attack missions, radar-equipped ADs became adept at attacking North Korean targets by night, especially in missions against the transportation network to disrupt the flow of supplies to the enemy's front line. Special versions of the Skyraider also flew electronic warfare and airborne early warning missions. Above is an AD-4NA of VF-54, seen in May 1952 between the squadron's two combat cruises. Note the hatch in the fuselage side for access to the rear crew compartment. The AD-4NA was an AD-4N with the night attack equipment removed. The aircraft at left is an AD-4N aboard Valley Forge, *complete with the antenna pod for the APS-19 radar under the starboard wing.*

Above and left: Two Grumman F6F-5K drones are prepared for launch from USS Boxer on 2 September 1952. These aircraft were unmanned and loaded with 1,000-lb (454-kg) bombs. They were 'flown' to their targets by a controller in an AD-2Q Skyraider. Six launches were undertaken by Guided Missile Unit 90, the first on 28 August. Aimed at bridges, tunnels and power plant, the attacks achieved two hits and a near-miss.

Right and below: Korea was the war in which the helicopter came of age. Sikorsky's S-51 was widely used by the Air Force (H-5) and Navy (HO3S-1). Utility Helicopter Squadron HU-1 provided detachments to carriers to act as plane-guards and to provide a limited liaison capability.

While by the time of the Korean War the Grumman TBM Avenger had been replaced in most of its front-line roles by the Skyraider and other types, there was still useful work for the veteran to perform. Its most important role was ASW, and sizeable numbers of TBM-3W2s and TBM-3Ss were deployed to the theatre. As the submarine threat failed to materialise, the VS squadrons were usually left ashore to make more room for attack aircraft. Below is a TBM-3S seen at Pohang in 1953. Avengers also undertook a variety of secondary roles. The line-up at left comprises the TBMs and F7F Tigercats of the Itazuke-based utility squadron, seen in February 1952. The aircraft are brightly painted for their role as target-tugs. Another important duty was COD (carrier onboard delivery) for which dedicated TBM-3Rs and stripped out TBM-3Ss were used. Avengers performed sterling work in evacuating casualties.

Above and left: USS Badoeng Strait (CVE 116) operated as an ASW carrier in 1951/52, carrying the TBM-3W2 'skimmers' of reservist unit VS-892 aboard. The APS-20 radar was intended to spot surfaced or snorkelling submarines for attack by TBM-3S2 'scrappers'.

Right: Dominating the scene aboard USS Badoeng Strait (CVE 116) in March 1952 is a PBM-5A. The 'Commencement Bay'-class carrier was ferrying aircraft back to the US for repair. The Martin PBM Mariner was a World War II-era flying boat that operated on maritime patrols throughout the Korean War. VP-47 was already in-theatre when the North invaded, and another six units (including Naval Reserve squadrons) were eventually involved in the conflict. Operations were normally staged from Oppama and Iwakuni in Japan, although Mariners were also regular visitors to Korean airfields. Missions lasted for up to 12 hours and were flown along the Korean and Chinese coastline.

Below: A PBM-5A, based at Oppama, sits on the ramp at Suwon (K-13) in 1953. The PBM-5A was the last production version of the Mariner, and had full amphibious capability.

Land-based maritime patrol was the province of the Consolidated PB4Y-2 Privateer (P4Y-2 from 1952). The covered aircraft above was seen at Iwakuni, while the VP-28 aircraft below was visiting Kimpo (K-14).

F-4E(S) Shablool

When an integrated Soviet air defence system was established in Egypt from 1970, Israel was faced with a major threat to its air superiority in the region. In order for the threat to be countered, intelligence had to be gathered concerning the whereabouts of the SAM sites – a task which only a high-flying aircraft with long-range cameras could perform safely. As a result, three IDF/AF Phantoms were modified to become F-4E (Specials).

Reconnaissance may not be a glamorous mission but in many cases it is much more dangerous than any other combat assignment. It is also a very demanding mission that takes the recce platforms to the limits of their flight envelopes. In many operational scenarios recce aircrew fly higher, faster and in a vastly more hostile environment than their colleagues.

A case for LOROP

Being such a high-value target on the one hand, and handicapped by the passive nature of the recce mission on the other, the reconnaissance aircraft naturally attracts enemy reaction. Since the early 1960s the Surface-to-Air Missile (SAM) has

been the main threat that endangered recce aircraft. The threat increased as SAM technology evolved, and as previously scattered SAM batteries were grouped to field an integrated air defence system (IADS). Flying higher and faster over IADS-defended sectors became more and more dangerous. Reconnaissance assets that could photograph enemy territory from above the ceiling of the IADS envelope became treasured strategic assets that only a few powers could develop and maintain. Strategic reconnaissance assets were therefore vastly expensive, relatively scarce and not always readily available – not even to the few powers that possessed such valuable national security assets in the late 1960s and early

1970s. The system that bridged the gap between tactical recce and strategic reconnaissance was the LOng Range Oblique Photography (LOROP) system.

Middle East monitor

The Israel Defense Force/Air Force (IDF/AF) also attached the highest priority to aerial reconnaissance, to the extent that McDonnell Douglas RF-4E Phantom II recce aircraft were included in the initial Israeli F-4E Phantom contract for 50 aircraft. Deliveries of the F-4E fighters commenced in September 1969 but the delivery of the RF-4Es was not expected before late 1970 or early 1971.

Israel's special reconnaissance Phantoms

Before and after – F-4E Block 44 Kurnass 146 (USAF 69-7567) is seen on display (left) with a variety of bombs at an air show in 1971. It was one of three similar aircraft from the Bat Squadron that were sent to the US for conversion to F-4E(S) Shablool standard, and is seen above wearing the tail number 498 that it carried throughout its reconnaissance career.

No diplomatic campaign to end the Arab-Israeli conflict followed the clear-cut Israeli victory in the June 1967 Six-Day War. To the contrary, Egypt's President Abdel Gamal El-Nasser stamped the slogan 'what had been taken by force will be returned by force', while in September 1967 Arab leaders adopted the 'three-no' policy at the Khartoum Summit: no recognition of Israel, no negotiation with Israel and no peace.

As originally delivered the three Shablool aircraft had a false black radome painted on to hide their identity from casual onlookers and aerial opposition. This is demonstrated on 492 above. However, Shablool 499 had its nose repainted in tan (below) in the early 1990s to commemorate the Bat Squadron's famous Mirage IIICJ(R) that was similarly painted during the 1970s.

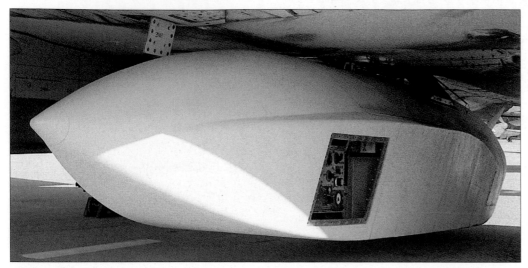

The G-139 pod carried by the RF-4E contained the HIAC-1 camera with large optical windows on both sides. The large size of the pod resulted in limited ground clearance, while its drag and weight penalty resulted in severe operational limitations.

In light of this diplomatic dead-end, limited-scale hostilities continued in the wake of the Six-Day War. Perhaps the most significant Arab lesson from the Six-Day War was that Arab fighter pilots using Soviet hardware and tactics were no match for the Israelis. The Egyptian and Syrian response to the challenge of IDF/AF air superiority was not in the air but on the ground. In 1968 the Egyptian Air Defence Force (ADF) was established to assume responsibility for the protection of Egyptian airspace. By the deployment of a network of SAM sites overlapping in coverage, protected by AAA and supported by control centres, radar stations and MiG-21 interceptors, the ADF was to challenge Israeli air superiority.

Backed by the USSR and confident in the capabilities of the restructured Egyptian armed forces, President Abdel Gamal El-Nasser launched the Attrition War that lasted from March 1969 through to August 1970. By early 1970 Israel had the upper hand in the static war along the 160-km (100-mile) Suez Canal. The Egyptian ADF crumbled while Israeli F-4E jets bombed strategic targets all over Egypt virtually unchallenged. Egypt's ability to wage a war of attrition was devastated, yet the Egyptian leadership refused to terminate hostilities without a military achievement. Countering the F-4Es was not a trivial matter, and the Egyptian response was indeed radical. Egypt's President Nasser visited Moscow in late January 1970 and persuaded his hosts that the only solution to avoid another humiliating defeat to a Soviet ally was direct Soviet intervention.

At the time there were already 1,500 Soviet advisors deployed to Egypt. In fact, Soviet advisors were posted to every Egyptian military outfit from regiment-sized units upwards and several were killed in the war, but in early 1970 a complete Soviet ADF Division, including an integral MiG-21 air brigade, was deployed to Egypt. Equipped with AAA, radar stations, command and control facilities, and the latest versions of the SA-2 and SA-3 SAMs, the Soviet Division initially assumed responsibility for the defence of Alexandria, Cairo and the Aswan Dam, thus freeing the Egyptian ADF to engage the IDF/AF in the battle for air superiority west of the Suez Canal.

Gradually the pendulum of warfare swung in favour of the Soviet-backed Egyptian ADF. By the summer of 1970 SAMs accounted for five Israeli F-4Es shot down with another two severely damaged. Overflying the Egyptian-Soviet IADS west of the Suez Canal became more and more dangerous at a time when near-real time intelligence data collection became critical to enable Suppression of Enemy Air Defence (SEAD) operations.

IADS supreme

The Attrition War ended in August 1970 with a US-brokered ceasefire. The success of the Egyptian-Soviet IADS actually enabled the Egyptian leadership to accept a ceasefire agreement. Yet the ceasefire was fragile and both sides

RF-4X mock-up

These three views show the mock-up of the RF-4X proposal in the General Dynamics works at Fort Worth in December 1974. IDF/AF F-4E 69-7576 was fitted with RF-4X features on the starboard side only. The most obvious was the mocked up tank above the engine trunk, which would have provided water injection for the PCC system. The intake was dramatically enlarged, as displayed in the head-on view, and featured a complex set of fully variable ramps, vortex generators and bleed ducts to allow tight control of the airflow up to a maximum dash speed of Mach 3.2. The nose was also recontoured (again on the starboard side only) to show how it would look with the necessary installation for the HIAC-1 camera. Although the F/RF-4X did not progress beyond this mock-up, the nose installation was used in the follow-on F-4E(S) and 69-7576 became the lead aircraft for the revised and less ambitious programme.

Left: The three Shablools were converted in General Dynamics' secretive Building 30, seen here in November 1975. The left-hand aircraft carries the US civil registration N97570, which was based on the aircraft's USAF serial and used for flight tests.

Below: The first F-4E(S) is rolled out at Fort Worth on 17 December 1975. The marks on the aircraft's nose are for a tufting layout so that airflow around the front end can be observed accurately.

prepared for the resumption of hostilities, as none of the basic issues that triggered the Attrition War in 1969 had been resolved. Israeli monitoring of the Egyptian ADF's deployment was therefore critical but could only be accomplished from afar, a classic case for LOROP.

The IDF/AF had over two decades of LOROP experience, having utilised modified transport aircraft to fly along the borders to 'peep' across the line. The initial IDF/AF LOROP platform was the ubiquitous Douglas C-47 equipped with a K-17 camera. The IDF/AF LOROP mission was handed over to the Nord 2501 Noratlas from the end of 1961. LOROP operations using transport aircraft were possible along 'peaceful' borders but unacceptable against an operational IADS. The IDF/AF required a higher resolution camera that would be able to fly high enough and far enough away to photograph the whole enemy's IADS from beyond the reach of the closest SAM batteries.

The RB-57F HIAC-1 option

The Six-Day War was not only a watershed in modern Middle East history but also in Israel's

The Peace Jack Phantoms were returned to 119 Squadron service during 1976. They acquired a fixed refuelling probe that complemented the standard boom receptacle soon after. 498 is seen here in its hardened aircraft shelter at Tel-Nof in 1978.

alliances. France was the principal supplier of arms to Israel until 1967 but from then on the USA took over. As a result of this shift, the evaluation by the Israelis of the Dassault Mirage IV for possible purchase as a recce platform was no longer viable. Looking at the US market for recce capabilities, the Israelis were faced with the tactical versus strategic issue. Tactical recce systems were readily available for export to Israel but could not challenge the IADS threat. Strategic recce systems such as the U-2 and SR-71 could but were not approved for export at all.

A possible exception to the US export policy was the RB-57F. The US Air Force purchased a

Martin-produced version of the British English Electric Canberra light bomber as the B-57 Night Intruder. The General Dynamics RB-57F was a re-engined and re-winged Martin B-57 that could lift more than twice the Lockheed U-2's payload to almost the same altitude. The 21 modified airframes served from 1963 until premature retirement in 1974 due to the discovery of cracks. To the Israelis the most promising capability of the RB-57F was its General Dynamics HIAC-1 LOROP camera. The 1680-mm (66-in) focal length HIAC-1 camera had a resolution of 0.25 m (0.8 ft) from a 35-km (21.7-mile) range that degraded to 1-m (3.3 ft) from 125 km (77,7 miles).

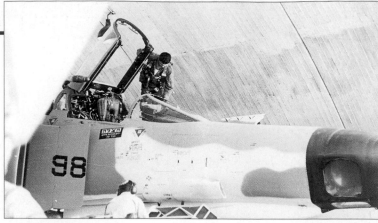

Above: 119 Squadron's original 499 was this RF-4E. When the Peace Jack aircraft arrived the tail number was transferred. In October 1976 this machine became 488 and was transferred to 69 Hammers Squadron.

Above right: For a short while during 1984 the Shablool jets had their last two digits painted in a large format on the air intake's splitter plates.

Right: Shablool 499 taxis with an AN/ALQ-119 jamming pod at Tel-Nof in December 1978.

Such an amazing capability was not achieved without a penalty and the trade-off was weight. The initial HIAC-1 mass was a staggering 1600 kg (3,527 lb).

Though the RB-57F was not as 'strategic' as the U-2 or SR-71, and although a single jet was once loaned to Pakistan, the Israeli request to purchase the HIAC-1 equipped RB-57F was denied. It was then that technology came to the aid of the desperate Israelis. The HIAC-1's weight dropped from the initial 1600 kg in 1963 to only 680 kg (1,500 lb) by 1971, thus enabling installation in platforms other than the RB-57F. The USAF exploited the technology leap to pack the HIAC-1 into the huge 1800-kg (3,968-lb) General Dynamics G-139 pod. Carried under the centreline station of the RF-4C, the G-139 could no longer be labelled 'strategic'. Yet it was an Israeli operational failure that highlighted the superiority of the Soviet IADS and actually triggered the supply of the G-139 pod to Israel.

Problematic pod

On 17 September 1971 an Egyptian show of force targeted an IDF/AF Boeing KC-97G Stratofreighter. The modified transport aircraft was on a routine LOROP mission to peer at the Egyptian military deployment to the west of the Suez Canal. Flying from north to south along the Suez Canal at an altitude of almost 30,000 ft (9144 m), the slow and large LOROP platform was well within the Israeli-occupied territory of Sinai. However, an inadequate camera specification coupled with the inability to fly higher and faster placed the KC-97G at the limits of the Egyptian ADF engagement envelope. Yet flying over Sinai was deemed safe, a belief that was shattered at roughly 1410hr local time just as the LOROP aircraft was half way along the reconnaissance run. For whatever reason, an Egyptian ADF SA-2 shot down the KC-97G. Only a single crewmember baled out as the aircraft spiralled down, the remaining seven aircrew being killed.

Cold War regional conflicts were test grounds for weapons and tactics, but also had the potential for global destabilisation. The shooting down of the LOROP KC-97G pointed at the superiority of the Soviet IADS but could, potentially, restart a regional conflict that a fragile US-brokered ceasefire had ended only a year before. Both issues were unacceptable in the view of US diplomats, so the G-139 HIAC-1 pod system was hurriedly delivered to Israel.

However, carrying the G-139 pod severely hampered the performance of the RF-4E. The added drag of the large pod restricted top speed to Mach 1.5, while the most critical restriction for recce operations was the resulting altitude limit of 50,000 ft (15240 m). At such altitude the HIAC-1 camera's advantage of high-quality, long-range photography was not exploited to the full. The obvious solution was an internal installation of the HIAC-1 in a high-flying jet. Luckily for Israel, the

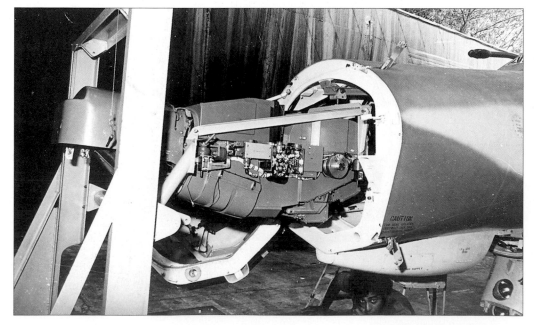

The HIAC-1 took up most of the Phantom's nose section, and slid forward for removal (left). The lens could be rotated to peer through oblique (above) windows to either side, or downwards. In the rear of the undernose bulge, just forward of the nosewheel, was a vertical KS-87 camera that was used to provide reference for the much more powerful HIAC-1.

Details of the Shablool's operational history are scant, but it is thought that it routinely undertook missions across Israel's borders. The only publicised mission involved a manoeuvre kill against a MiG-21 being logged during an operation into Iraq.

HIAC-1's weight dropped by then to actually enable such an installation, while a promising platform was under development to fulfill an entirely different requirement.

Converging requirements

Israel was unable to overfly the Egyptian IADS but the Egyptians had access to intelligence data collected by Soviet MiG-25 recce jets that were based in Egypt from March 1971. IDF/AF F-4E aircrews attempting to intercept the high and fast MiG-25s only managed to record a few close calls. Clearly an interceptor superior to the F-4E was required to intercept the MiG-25, but none existed at the time. The only available option was to improve the F-4E's climb and speed performance, and the Israeli Technion Institute of Technology's Jet Engines Laboratory studied the application of Pre-Compressor Cooling (PCC) to increase the General Electric J79 turbojet's thrust, thus enhancing the F-4E's performance.

The MiG-25 challenge was not unique to Israel. The USAF financed General Dynamics to develop a PCC Phantom, resulting in the F-4X project. PCC necessitated an extensive air intake modification and the addition of two water tanks. The

tanks were bolted to the airframe on top of the fuselage. Each tank had a capacity of 1135 kg (2,500 lb) of water, as well as being a self-contained system that included all the necessary pumps to avoid additional radical structural modifications. Water sprayed into the air inlets increased the mass airflow with a calculated 150 per cent thrust augmentation at high speed and

high altitude. Though PCC added drag and weight, the performance gains were expected to be significant. The F-4X was designed to dash at Mach 3.2 and cruise at Mach 2.4. The exceptional performance of the F-4X might have enabled the PCC Phantom to intercept the recce MiG-25 but also threatened a major USAF project that was also designed to counter the MiG-25. The McDonnell

Above: The pilot had a sight on the cockpit rail for aiming the HIAC-1 camera.

Right: This is the rear cockpit of F-4E(S) 492. The main instrument panel was generally standard for an F-4E, with threat warning screen at the top right. The screen display below the panel replaced the radar display of the standard fighter.

Douglas F-15 Eagle was only two to three years away from USAF service entry at the time of F-4X development. Moreover, in the wake of the October 1973 Yom Kippur War, the US Ford Administration committed to supply the F-15 to Israel with IDF/AF service entry scheduled for 1975. The case for the F-4X 'MiG-25 hunter' had evaporated, but even the F-15 could not cruise at Mach 2.4 or dash at Mach 3.2. Clearly the PCC Phantom had the potential to be developed into a superb reconnaissance platform.

From F-4X to RF-4X

General Dynamics identified the potential of the F-4X and forwarded to the USAF on 12 April 1973 a proposal to develop a reconnaissance version of the PCC Phantom. The principal sensor was the HIAC-1 LOROP system that was accommodated inside a modified nose section from which the radar and cannon were removed. The RF-4X was designed to cruise at Mach 2.4

while flying at 78,000 ft (23774 m). Water capacity restricted the high-altitude, high-speed LOROP run to just 10 minutes. The pilot manually activated the water injection system while a visual indication notified the pilot as water supply ran low. An automatic system decreased the fuel flow to the engines just as water supply ended to avoid potential damage to the J79 turbojets.

The modified nose section was longer and bulkier, corresponding with a 30-cm (12-in) increase in the jet's overall length. Internal nose section volume was 1.98 m³ (70 cu ft). An environment control system was designed to maintain a selected nose section temperature at a tolerance of only ±2° Celsius. The HIAC-1 was mounted on a swivelling mount and was pointed at a target by an optical sight on the canopy sill. Four rectangular windows, two side windows and two at the bottom of the nose section were provided for the HIAC-1, while a fifth circular window served a vertical KS-87 camera that was located behind the

HIAC-1's installation. As an alternative, a CAI KA-90 camera could have been installed in place of the HIAC-1. With the HIAC-1 installed, the RF-4X was expected to cover a 400-km (250-mile) run during a four-minute, maximum-altitude, maximum-speed dash. The images produced covered an area of more than 60,000 km².

An IDF/AF 119 'Bat' Squadron F-4E Kurnass (Sledgehammer) tail/number (t/n) 167 was delivered to General Dynamics on 6 November 1974 to serve as the RF-4X pattern aircraft. By the next month this jet was already modified as an RF-4X mock-up with representative models of a water tank, a larger nozzle and a modified nose section sculptured onto the right-hand side of the jet, with the left-hand side retaining the F-4E look. However, the project did not progress beyond that modest landmark. As its exceptional performances pushed the jet into the 'strategic' domain, the US was reluctant to export the RF-4X. Combined with the USAF's commitment to the

Shablool aircrews had to wear David Clark pressure suits on LOROP missions that were flown at altitudes of more than 50,000 ft (15240 m). The $120,000 suit weighed more than 10 kg (22 lb) and took 20 minutes to put on. Each pilot had an individually tailored suit, stored in the Bat Squadron's 'High-Altitude' room (right).

Two views show Shablool 492 in its element, highlighting the refuelling probe fixture that ducted fuel into the fuel system just forward of the boom receptacle. For the best photographic results the aircraft flew high and fast, and the Peace Jack aircraft routinely operated at altitudes above 60,000 ft (18288 m). The highest altitude recorded by an IDF/AF reconnaissance Phantom was 73,400 ft (22372 m).

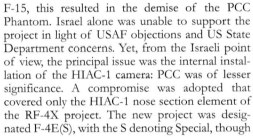

F-15, this resulted in the demise of the PCC Phantom. Israel alone was unable to support the project in light of USAF objections and US State Department concerns. Yet, from the Israeli point of view, the principal issue was the internal installation of the HIAC-1 camera: PCC was of lesser significance. A compromise was adopted that covered only the HIAC-1 nose section element of the RF-4X project. The new project was designated F-4E(S), with the S denoting Special, though in Israeli service the new Phantom variant was to be known as the Shablool (Snail).

'Niners' legacy

Israel assigned two additional 119 Bat Squadron F-4E Kurnass jets to the Peace Jack F-4E(S) conversion project. All three jets had flown combat missions in 119 Bat Squadron service – one was a MiG-killer and another was delivered with patches covering battle damage holes. General Dynamics rolled out the first F-4E(S) on 17 December 1975 with first flight following on 20 December. The first Peace Jack jet was fitted with wool tufts all around the modified nose section to record the airflow during flight-testing. Once General Dynamics completed the flight envelope and system functionality tests, the three Shablool recce jets were delivered to Israel and entered service with the Tel-Nof based 119 Bat Squadron during 1976.

The Bat Squadron had been the premier Israeli recce squadron since 1964. The squadron operated the two unique Dassault Mirage IIICJ(R) recce birds that were tail-numbered 98 and 99. When an RF-4E flight was established in the Bat Squadron in 1971, the famous tail numbers were assigned to two of the new Phantoms. By then,

Shablool 492 lands at Tel-Nof in 1990. The F-4E(S) Phantoms were equipped with the so-called 'hard wing' without manoeuvre slats. The wing is much cleaner aerodynamically, an attribute that could really be appreciated at over Mach 1.5.

Refuelling played a prominent part in F-4E(S) operations, the aircraft having a probe for tanking from KC-130s and, as here, a receptacle for refuelling from Boeing 707 Re'em tankers. In the background another Re'em refuels an F-15, which often provided ingress/egress escort for Shablool missions. The F-4E(S) is carrying a Sidewinder missile on a launch rail mounted in the forward Sparrow bay.

however, the IDF/AF had shifted to a three-digit tail number system so the two Bat Squadron prestigious recce birds were actually 198 and 199. A further revision of the IDF/AF tail number system resulted in all RF-4Es changing the prefix digit from 1 to 4 immediately after the October 1973 Yom Kippur War, so the two top RF-4E 'niners' became 498 and 499.

Brigadier General Amir Eshel and Brigadier General Yochanan Locker depart Tel-Nof on 17 May 2004 to deliver Shablool 498 to the IDF/AF Museum.

The F-4E(S) service entry resulted in a rare IDF/AF re-serial numbering case. Bat Squadron 119 was so attached to the tradition of operating the '98' and '99' recce platforms that the IDF/AF staff was persuaded to allocate the serial numbers to two of the newly delivered Shablool jets. F-4E(S) 69-7567 became Shablool 498 and F-4E(S) 69-7576 became Shablool 499. The two original '98' and '99' RF-4Es were re-numbered 488 and 489, respectively, and were transferred from Bat Squadron 119 to Hammers Squadron 69. The third F-4E(S), 69-7570, became Shablool 492 to fill a gap in the recce 'niners' t/n sequence.

Classified operations

The nature of aerial reconnaissance is such that a low profile is essential to maintain a successful operation. Only from time to time the exploits of reconaissance aircraft are released to the public,

and the F-4(S) jets managed to keep an extremely low profile. One bizarre exception occurred on 3 January 1982 when a MiG-21 crashed while attempting to intercept Shablool 498 flown by Bat Squadron 119 commander Lieutenant Colonel Gideon Sheffer with Captain Yuval Naveh as his navigator, and Shablool 492 flown by Major Ran Granot and Captain Daniel Grossman. The pair of recce Phantoms was on a reconnaissance mission over Iraq, and was intercepted at low altitude after the successful completion of the high-altitude photo run.

Lieutenant Colonel Sheffer quickly analysed the situation as the lone MiG-21 closed on the two Phantoms. The MiG-21 was flying with full afterburner in a clean configuration and had been probably doing so since it was scrambled from its base – in other words its fuel was rapidly running out. The two recce Phantoms engaged the MiG-21 for a few minutes, ensuring its destruction without a single shot being fired. Moreover, the MiG-21 only had a tail-chase air-to-air missile (AAM) option while the recce Phantoms had the all-aspect Rafael Python 3 AAM for self-protection. At any given moment of the engagement the recce Phantoms denied the MiG-21 an AAM launch solution, while at the same time retained the option to turn into the MiG and launch an AAM at off-boresight angle.

As much as the opportunity to shoot down a MiG-21 looked attractive, the Shablool crews were focused on their mission: to bring home the precious films while keeping the profile of their mission as low as possible. The logical decision was made and the two recce Phantoms disengaged in a manoeuvre rather than using their superior firepower. Racing back home to the waiting A-4 Skyhawk buddy tankers, and already escorted by F-15 Eagles, the Shablool crews could only guess that the MiG-21 was already doomed as it

All three Shablools were retired on 12 May 2004 – here are the last operational landings of 498 (above) and 492 (right). By this time they had been transferred from 119 Bat Squadron to the One Squadron (201) and wore the unit's well-known chevron markings. 498 sports an Iraqi kill marking on its nose for the 1982 'manoeuvre' kill, while 492 has an Iraqi roundel for a kill dating from its days as a standard F-4E Kurnass.

exhausted its precious fuel during the engagement.

An intelligence report confirmed the fact that the Iraqi MiG-21 indeed crashed in the desert due to fuel starvation, though the pilot ejected. The kill was credited to the Shablool 498 crew and was labelled a 'manoeuvre' kill, a term that replaced the 'no weapon' label that resulted in a squadron kill credit rather then in a personal kill credit as the 'manoeuvre' label entitled. Shablool 498's kill was almost the last Israeli Phantom air-to-air kill, the last air-to-air kill for Bat Squadron 119 thus far and the only air-to-air kill credited to an Israeli recce Phantom.

Shablool twilight

The three Shablool recce birds provided Israel with an exceptional Intelligence Surveillance and Reconnaissance (ISR) capability for almost three decades. In 1988 Israel launched the first of the Ofeq (Horizon) series of Low Earth Orbit (LEO) reconnaissance satellites, while real-time recce pods were introduced to equip the IDF/AF F-15 and F-16 combat aircraft. All of these new ISR assets produce digital images that are transmitted to a ground station via a secured broadband datalink. Yet the Shablool's non-digital and non-

real-time HIAC-1 camera produced image quality that only a few modern ISR systems could match. It was the Shablool's F-4 legacy platform that actually dictated the retirement of the Shablool from Israeli service rather than the sensor system. The final chapter in the Shablool's career was written on 17 May 2004 when the then Tel-Nof air base commander, Brigadier General Amir Eshel, with the then Hatzerim air base commander, Brigadier General Yochanan Locker, in the back seat ferried Shablool 498 from Tel-Nof to Hatzerim, where the unique jet has been placed on display at the IDF/AF Museum.

Shlomo Aloni

498 is now on display in the IDF/AF museum at Hatzerim. Immediately after delivery it was returned to Bat Squadron markings to honour the unit with which it flew for most of its career.

Picture acknowledgments

Front cover: Rick Llinares/Dash 2, Ted Carlson/Fotodynamics (two). **4:** Lockheed Martin via David Donald. **5:** Eurofighter via David Donald. **6:** Saab via David Donald (three). **7:** via Tom Kaminski (two). **8:** Bell via David Donald. **9:** via Tom Kaminski, Jonathan Chuck. **10:** USAF via David Donald, Lockheed Martin via David Donald. **11:** Shlomo Aloni, via Tom Kaminski. **12:** Saab via David Donald, US Navy via Tom Kaminski. **13:** Chris Lofting, Eric Katerberg, via Tom Kaminski. **14:** Erich Strobl, Peter R. Foster. **15:** via Tom Kaminski, Shlomo Aloni. **16-21:** Rick Llinares/Dash 2. **22:** Dick Lohuis (three). **23:** via Terry Panopalis (two), Terry Panopalis. **24:** USMC via David Donald. **25:** Sikorsky via David Donald. **26:** AWI via Tom Kaminski, EADS via Tom Kaminski. **27:** MDH via Tom Kaminski, Bell via Tom Kaminski. **28:** US Army via Tom Kaminski (two). **29:** Alenia via Tom Kaminski, EADS via Tom Kaminski. **30-31:** Saab via David Donald. **32-33:** BAE Systems via David Donald. **34:** via Nigel Pittaway (two). **35:** via Nigel Pittaway, Nigel Pittaway, Boeing via David Donald. **36:** via Nigel Pittaway (three). **37:** via Nigel Pittaway (two), Nigel Pittaway. **38:** Nigel Pittaway (nine). **39:** via Nigel Pittaway, Nigel Pittaway (two). **40:** Nigel Pittaway (three), via Nigel Pittaway. **41:** via Nigel Pittaway (two), Nigel Pittaway. **42:** via Nigel Pittaway (two). **43:** via Nigel Pittaway, Boeing via David Donald. **44:** René van Woezik (two), Cees-Jan van der Ende. **45:** René van Woezik (two). **46:** René van Woezik (two), Cees-Jan van der Ende. **47:** Cees-Jan van der Ende, René van Woezik (two). **48:** René van Woezik (two), Cees-Jan van der Ende. **49:** Cees-Jan van der Ende, René van Woezik. **50:** Gert Kromhout, US Navy via Gert Kromhout. **51:** US Navy via Gert Kromhout. **52:** US Navy via Gert Kromhout (two). **53:** US Navy via Gert Kromhout, Gert Kromhout. **54:** US Navy via Gert Kromhout, Gert Kromhout. **55:** US Navy via Gert Kromhout, Gert Kromhout. **56:** US Navy via Gert Kromhout (three), Gert Kromhout. **57:** US Navy via Gert Kromhout, Gert Kromhout. **58:** Gert Kromhout (four). **59:** Gert Kromhout (three), US Navy via Gert Kromhout. **60-61:** Ted Carlson/Fotodynamics. **62:** Ted Carlson/Fotodynamics, USMC via David Donald. **63:** Ted Carlson/Fotodynamics, US Navy via Peter Cline via Gert Kromhout. **64:** Gert Kromhout, US Navy via David Donald, USMC via David Donald. **67:** Ted Carlson/Fotodynamics (three), Gert Kromhout (two). **68:** USMC via David Donald (two), Gert Kromhout. **69:** US Navy via David Donald (two), Ted Carlson/Fotodynamics. **70:** US Navy via Gert Kromhout (two), US Navy via Brad Elward. **71:** Ted Carlson/Fotodynamics, US Navy via Brad Elward. **72:** LTCOL Kevin Gross USMC via Gert Kromhout. **73:** Ted Carlson/Fotodynamics. **74:** Gert Kromhout (two). **75:** Ted Carlson/Fotodynamics, LTCOL Kevin Gross USMC via Gert Kromhout. **76:** USMC via Gert Kromhout, Ted Carlson/Fotodynamics. **77:** Ted Carlson/Fotodynamics, US Navy via David Donald. **78:** USAF via David Donald, USMC via David Donald. **79:** USMC via David Donald (two), USAF via David Donald. **80-81:** USAF via David Donald (four). **82:** US Navy via David Donald, Ted Carlson/Fotodynamics. **83:** Ted Carlson/Fotodynamics, USMC via David Donald. **84:** Cees-Jan van der Ende (two). **85:** Luigino Caliaro, Cees-Jan van der Ende (two). **86-89:** Cees-Jan van der Ende. **90-97:** Ted Carlson/Fotodynamics. **98:** Ted Carlson/Fotodynamics (three), USAF via David Donald. **99-101:** Ted Carlson/Fotodynamics. **102-107:** Stefan Degraef and Edwin Borremans. **108:** Aerospace (two). **109:** Aerospace, Terry Panopalis. **110:** Terry Panopalis Collection, Barry Jones Collection via Terry Panopalis. **111:** Terry Panopalis Collection, Barry Jones Collection via Terry Panopalis. **112:** Terry Panopalis Collection, Aerospace. **113-117:** Aerospace. **118:** Aerospace, Terry Panopalis Collection. **119:** Terry Panopalis Collection (two). **120:** Barry Jones Collection via Terry Panopalis, Aerospace (two). **121:** Barry Jones Collection via Terry Panopalis (two), Aerospace (four). **122-123:** Aerospace. **124:** Aerospace, Barry Jones Collection via Terry Panopalis. **125:** Terry Panopalis Collection. **126:** Dr Alfred Price, Aerospace. **127-128:** Aerospace. **129:** Aerospace, Dr Alfred Price (three). **130:** Aerospace, Dr Alfred Price (two). **131:** Aerospace, Dr Alfred Price (three). **132:** Dr Alfred Price (two). **133:** Dr Alfred Price, Aerospace. **134:** Aerospace (two), Dr Alfred Price (two). **135:** Aerospace, Dr Alfred Price (two). **136:** Dr Alfred Price (two), Aerospace (two). **137:** Dr Alfred Price, Aerospace (three). **138:** Aerospace, Dr Alfred Price (three). **139:** Dr Alfred Price (two), Aerospace. **140:** Aerospace (nine). **143:** Aerospace (three). **144:** Aerospace (three), Dr Alfred Price. **145:** Aerospace (six), Dr Alfred Price. **146:** Dr Alfred Price (three), Aerospace. **147:** Aerospace (two). **148:** Dr Alfred Price (three). **149:** Aerospace (four), Dr Alfred Price. **150:** Dr Alfred Price (two), Aerospace. **151:** Aerospace (five), Dr Alfred Price (six). **152:** Dr Alfred Price (four). **153:** Aerospace (two), Dr Alfred Price. **154:** Dr Alfred Price, Aerospace (three). **155:** Dr Alfred Price (three), Aerospace (three). **156:** John Leverton via Warren Thompson via WT, via WT (two). **157:** Bruce Bagwell via WT, Bob Balser via WT (two), Tom Blackburn via WT. **158:** Gene Bezore via WT, E.H. Yeager via WT, Ben Sowaske via WT, Len Gordinier via WT, Cale Herry via WT. **159:** Frank Jones via WT, John Corrigan via WT, Norman Green via WT. **160:** Don Frazor via WT (three), Bob Balser via WT. **161:** Dick Staringchak via WT (two), Hal Schwan via WT, Frank Jones via WT. **162:** Harry Winberg via WT, Pierkowski via WT, Tom Archer via WT, Gerald Haddock via WT (two). **163:** Gerald Haddock via WT, Robert Daniels via WT, Van Shaghoian via WT, Earl Plesia via WT. **164-173:** Shlomo Aloni.